The Architecture of Encounter

ARCHITECTURE OF ENCOUNTER

The Architecture of Encounter

A Mediated Encounter Ontology

Bry Willis

Philosophics Press ◐ 2026

First Edition

ISBN (Hardback): 978-1-972025-02-4
ISBN (Paperback): 978-1-972025-03-1

ORCiD: 0009-0005-0016-5694

Published by *Philosophics Press*,
Cambridge, Massachusetts
Printed in the country of distribution
< philosophics.blog >

Library of Congress Control Number (LCCN): 2026906470

Subjects:

Ontology
Metaphysics
Philosophy of language
Knowledge, theory of

LCC: BD111

rev. a

to deflating mysticism from metaphysics

Education is an admirable thing, but it is well to remember from time to time that nothing that is worth knowing can be taught.

— Oscar Wilde

Preface

Why architecture

This book constructs a framework. It does not defend a thesis in the usual sense – advancing a single claim against objections and concluding with a vindication. It builds something: a set of structural commitments about how reality is organised, how agents are constituted, how knowledge is stabilised, how language coordinates and distorts, and how institutions inherit the limitations of the protocols through which they operate. The result is not a single argument but an architecture – a relational ontology that takes structured interaction as its primitive and derives the familiar categories of philosophy as stabilisations within it.

The method is diagnostic rather than prescriptive. The framework does not tell the reader what to believe about objects, minds, or institutions. It identifies how inherited conceptual structures – substance ontology, the mind-world bifurcation, the noumenal-phenomenal split – generate problems that are then mistaken for deep features of reality, and it offers an alternative structural vocabulary in which those problems either dissolve or change shape. The ambition is not to replace one set of doctrines with another but to clear conceptual space – to make visible the architectural assumptions that philosophy has been building on, largely unexamined, and to show what happens when different assumptions are installed.

The approach is compositional rather than polemical. Each chapter develops a component of the architecture – encounter, constraint, subjecthood, logic, epistemology, perception, language, social ontology, realism – and connects it to the others through shared structural commitments. The chapters are designed to be self-contained: a reader can enter at any point and follow the argument without having read what precedes it, though the cumulative effect of reading in sequence is, inevitably, stronger. The architecture is meant to be assessed as a whole, not defended piecemeal.

The point of departure is the Mediated Encounter Ontology of the World – MEOW – which I developed in an essay published independently (Willis, 2025). That essay proposed that encounter-events are ontologically primary, that 'mind' and 'world' are derivative abstractions, and that mediation is constitutive rather than obstructive. The present work takes that essay as a starting point and extends it substantially.

'Extends' is precise. Beyond MEOW does not merely restate the original framework with additional commentary. It revises formulations that were imprecise or underspecified in the original essay – most notably the account of subjecthood, which has been reconceived as modal differentiation rather than geometric coordinate. It integrates components that were absent from the original – affordances, ontological grammar, the Language Insufficiency Hypothesis, social ontology as second-order constraint. It strengthens commitments that were asserted but not defended – particularly the realist anchor, which now rests on an explicit convergence argument rather than a promissory note. And it identifies limits that the original essay did not acknowledge – above all, the normativity frontier.

The original essay is reproduced in Appendix C, without modification. The reader who wishes to compare the monograph's treatment with its point of departure may do so. The comparison is invited, not feared. A framework that cannot survive revision is not an architecture. It is a monument.

On the status of claims

This book makes several kinds of claim. They are not the same kind, and they should not be evaluated as though they were.

Some claims are inherited. The framework draws on established results from physics, cognitive science, evolutionary biology, and the philosophy of perception. These are not argued for here. They are adopted as premises, cited as such, and subject to revision if the sciences that produced them revise them. The reader who wishes to contest them should take that up with the relevant disciplines, not with this book.

Some claims are architectural. The ontological primacy of encounter-events, the derivative status of mind and world, the account of ontological grammar as a mechanism of reification, the treatment of coherence zones as heuristic stabilisations rather than natural kinds – these are the framework's own proposals. They are defended on grounds of explanatory power, internal coherence, and comparative adequacy. They do not require direct experimental confirmation. They require that they do the work better than the structures they replace.

Some claims are diagnostic. The hard problem of consciousness as a nominalisation misfire, the realism-idealism impasse as an artefact of the subject-object architecture, the institution as a second-order constraint system – these are applications of the architecture to inherited problems. They stand or fall by whether they dissolve those problems more effectively than the frameworks that generated them.

And some things remain open. The normativity frontier is identified, not closed. The formal modelling of certain claims is sketched, not completed. The empirical investigation of reflexive integration is indicated, not performed. These are not hidden failures. They are visible borders.

The reader who wishes to object to a claim is invited to identify first which kind of claim it is. The appropriate standard of evaluation is not the same in each case. A framework that had to answer all four categories by the same criterion would be either dishonest or incoherent – possibly both.

A word about what this book is not

It is not a contribution to cognitive science, though it draws on and engages with work in ecological psychology, enactivism, and predictive processing. The claims made here are philosophical – they concern the structure of encounter, not the mechanisms of neural processing. Where neuroscientific frameworks are discussed, they are reinterpreted within the architecture's commitments, not borrowed wholesale.

It is not a work in the philosophy of AI, though the AI-mind debate serves as a controlled demonstration in Chapter 10. The framework's application to that debate is diagnostic: it identifies the structural confusions that make the debate intractable. It does not resolve the question of machine consciousness and does not claim to.

It is not a political theory, though the account of institutional coordination and failure in Chapter 8 has implications for governance, policy, and democratic deliberation. Those implications are identified but not developed. The framework is diagnostic, not prescriptive. It explains why institutions built from language behave as they do. It does not tell them how to behave differently.

And it is not a closed system. The architecture identifies its open frontiers – normativity, the formal modelling of the effectiveness gradient, the empirical investigation of reflexive integration – and treats them as invitations rather than failures. A framework that claimed completeness would be, by its own diagnostic criteria, a system that had mistaken its conceptual ambitions for structural discoveries.

A note on the status of claims. The claims in this monograph fall into three categories: (1) empirical premises adopted from established sciences – these are not argued for but cited and relied upon; (2) architectural commitments proposed by the framework – these are the monograph's own proposals and are defended on structural grounds; and (3) diagnostic interpretations of philosophical problems that follow from those commitments – these are applications, testable by their explanatory purchase. The reader who wishes to contest a claim is invited to identify which category it belongs to, since the appropriate standard of evaluation differs in each case. The framework introduces a number of technical terms – encounter-event, modal differentiation, coherence grammar, ontological grammar, synchronisation protocol, second-order constraint system, institutional false convergence, convergence realism – that may be unfamiliar even to readers with philosophical training. These terms are defined where they are introduced and summarised in Appendix A. They are not coined for novelty. Each names a structural feature that existing vocabulary either does not capture or captures only by importing substance-ontological commitments the framework rejects. The terminological cumbersomeness is deliberate: it is the cost of refusing a lossy compression, and the reader is asked to judge whether the fidelity the terms purchase is worth the fluency they sacrifice.

A note on voice. This monograph is written in an academic register – the argument is developed through sustained prose rather than aphorism, and the engagement with other traditions is explicit rather than allusive. But it does not adopt the hedging, qualification-stacking, and defensive footnoting that characterises much contemporary analytic philosophy. The claims are stated plainly. The commitments are visible. Where the framework is uncertain, the uncertainty is named. Where it is confident, the confidence is not disguised as tentativeness. The reader is treated as a colleague, not as a tribunal.

The book that follows is an invitation to think structurally about problems that have been thought about substantively for too long. Whether the architecture holds is for the reader to determine. What the book asks is simple: examine the structure, follow its consequences, and decide whether the problems it dissolves were ever the ones philosophy needed to solve.

– Bry Willis, March 2026

Author Position Statement

Every philosophical framework carries commitments that precede its arguments. Some of these are methodological – preferences about what counts as a good explanation, what level of precision is worth pursuing, what kinds of evidence are admissible. Others are dispositional – habits of attention, trained suspicions, instinctive sympathies that shape which problems look urgent and which look manufactured. Honesty requires that these commitments be declared rather than concealed, so that the reader can assess the architecture on its own terms rather than suspect it of hiding its motivations.

What follows is that declaration.

I am a philosopher of language. My intellectual formation has been shaped by a persistent conviction that language is afforded undeserved authority in philosophical and institutional discourse – that it is routinely treated as a transparent medium for thought when it is, in fact, a structurally biased encoding protocol that shapes what can be fluently articulated and what cannot. This conviction did not begin as a theoretical commitment. It began as an observation: that many philosophical disputes which appear to concern deep features of reality are better understood as artefacts of the grammatical structures used to describe them. The observation has hardened, over time, into a methodological principle – one that this monograph applies across ontology, epistemology, perception, social theory, and the philosophy of mind.

I am sceptical of substance ontology. This scepticism is not a conclusion I arrived at after careful deliberation among the options. It is a starting disposition – a trained suspicion that nouns are doing more ontological work than they deserve, and that the philosophical tradition's commitment to things-bearing-properties reflects the grammar of the languages in which philosophy has been conducted rather than the structure of what is being described. I have tried, throughout this monograph, to argue for this scepticism rather than merely assert it. But the reader should know that the scepticism came first and the arguments followed. This is, I suspect, how most philosophical commitments actually work. The difference is that I am saying so.

I favour deflationary accounts of metaphysical categories. Where the tradition sees a deep ontological distinction – between mind and world, between fact and value, between substance and property – I tend to see a heuristic that has been promoted beyond its station. This preference is not nihilism. The categories are useful. Many are indispensable. But usefulness does not entail ontological depth, and the habit of treating convenient labels as discoveries about the furniture of reality is, in my view, the source of a significant proportion of philosophy's intractable problems. The junk drawer is full of real things. The drawer itself is not a natural kind.

I am trained as an economist as well as a philosopher, and the economic habit of mind – attending to constraints, trade-offs, compression, efficiency, and the structural conditions under which coordination succeeds or fails – runs through this monograph in ways that a reader from a purely philosophical background may find unfamiliar. The effectiveness gradient of Chapter 7 borrows its structural logic from the Production Possibility Frontier. The account of institutional coordination in Chapter 8 treats institutions as constraint-management systems rather than as expressions of collective will. The normativity frontier of Chapter 11 is framed, in part, as a question about what economic and institutional reasoning cannot reach. These are not analogies. They are structural debts to a discipline that understands constraint better than most branches of philosophy do.

The framework developed here draws on several traditions without belonging fully to any of them. The ecological psychology of Gibson, the enactivist work of Maturana and Varela, the inferential pragmatism of Sellars and Brandom, and the process metaphysics of Whitehead each contributed structural insights that the architecture absorbs and reconfigures. Bergson's diagnostic of spatialised thinking informs the ontological grammar thesis, though his epistemological remedy is rejected. The architecture departs from each of these traditions in ways discussed in the chapters that follow and detailed in Appendix B. The aim is not synthesis for its own sake but structural adequacy – taking what each tradition gets right and leaving behind the commitments that the framework's own diagnostics identify as problematic.

I hold several methodological commitments that have shaped the monograph's construction.

The first is that a framework is better judged by its structural consequences than by its intuitive plausibility. Substance ontology is intuitively plausible – nouns feel like they name things, and things feel like they exist independently. The intuition is a product of ontological grammar, not a guide to metaphysical truth. Throughout this monograph, I have prioritised structural fidelity over intuitive comfort, accepting the fluency tax that this imposes and asking the reader to do the same.

The second is that intellectual honesty about the limits of a framework is more valuable than the appearance of completeness. The normativity frontier is real. The formal modelling of the effectiveness gradient is incomplete. The substrate-neutrality of modal differentiation is a principled commitment, not an empirical demonstration. I have stated each of these openly because a framework that cannot acknowledge what it does not know is a framework that cannot be trusted with what it claims to know.

The third is that the framework is offered as architecture rather than doctrine. It is a set of structural commitments that can be tested, extended, revised, or abandoned. It is not a revelation. It is not a manifesto. It is a diagnostic instrument – a way of making visible the structural assumptions that philosophical and institutional reasoning operate on, usually without knowing it. The reader is invited to use the instrument, not to worship it.

A final disclosure. This monograph was developed in sustained dialogue with artificial intelligence systems – primarily Claude (Opus 4.6; Anthropic) and ChatGPT (5.2, Thinking; OpenAI), incidentally Gemini (3; Google) and Grok (4.1, Expert; xAI). These systems served as interlocutors, critics, and structural stress-testers throughout the writing process. They did not generate the framework. They did not supply its commitments. They did, at various points, identify weaknesses, propose refinements, and pressure-test arguments in ways that improved the final product. I mention this not because it is unusual – it will not be unusual for long – but because a monograph that diagnoses the structural limitations of language as a coordination protocol should be transparent about the coordination protocols it employed in its own production.

The ideas are mine. The architecture is mine. The errors, inevitably, are also mine.

Table of Contents

FOUNDATIONS

I

1

Mediated Encounter Ontology

Restating and refining the core framework

Radical empiricism takes conjunctive relations at their face-value, holding them to be as real as the terms united by them.

— William James, *Essays in Radical Empiricism*

To see a world in a grain of sand

— William Blake, *Auguries of Innocence*

1.1 Introduction: The Inheritance

Philosophy has long organised its inquiry around a foundational picture: a knowing subject set against a known world, with the central question being how the one gains access to the other. Whether framed as the problem of perception, the riddle of consciousness, or the dispute between realism and idealism, the underlying architecture remains constant. There are two domains – mind and world, inner and outer, subject and object – and the task of philosophy is to explain how they relate. This picture is not a discovery. It is an inheritance.

The Mediated Encounter Ontology of the World begins from the recognition that this inherited architecture generates more problems than it resolves. The realism–idealism dispute, the hard problem of consciousness, the location of meaning, the boundary between self and environment, the question of whether machines can think – each of these is sustained, at least in part, by the assumption that 'mind' and 'world' are independently existing domains that subsequently enter into relation. MEOW proposes that this assumption is the source of the difficulty rather than a constraint on its solution.

The present chapter restates and refines the core commitments of the framework. It is written to be self-contained: readers unfamiliar with the original essay need not consult it to follow the argument developed here and in subsequent chapters.[1] Where the present formulation departs from the original, the departure is noted explicitly.

1.2 The Impasse

The realism–idealism dispute has reached a stage of diminishing returns. This is not because the arguments on either side have been exhausted, but because both traditions draw from a common well – and the well is dry.

Realism maintains that thought and experience answer to a mind-independent world. The claim has obvious appeal: it underwrites scientific practice, common sense, and the intuition that reality does not bend to preference. But realism faces a structural difficulty it has never satisfactorily resolved. If all epistemic access is mediated – through perceptual systems, cognitive architecture, conceptual frameworks – then the claim to describe

1 The original essay, 'The Mediated Encounter Ontology of the World: A Relational Metaphysics Beyond Mind and World' (Willis, 2025), is reproduced in full as Appendix C.

a mind-independent reality is made from within the very mediation it purports to transcend. The realist either weakens the claim (admitting that we know reality only as mediated, which concedes the ground) or insists on a noumenal substrate accessible in principle but never in practice (which explains nothing). Realism oscillates between modesty and dogmatism, and neither posture is stable.

Idealism reverses the polarity. If all we ever encounter is structured experience, then perhaps experience or mind is the more fundamental category. The appeal is equally obvious: it takes seriously the fact that reality, as we know it, is always already shaped by the conditions of knowing. But idealism struggles with an equally stubborn difficulty. Experience is not infinitely plastic. Certain patterns persist regardless of how they are framed: gravitational effects that ignore conceptual preference, toxic substances that remain toxic despite linguistic reframing, perceptual limitations that resist training or will. Idealism must either accommodate this resistance (which pulls it toward the realism it set out to replace) or deny it (which costs it explanatory scope).

The oscillation between these poles is not evidence of insufficient ingenuity on either side. It is a symptom of a shared presupposition. Both realism and idealism assume that mind and world are independently existing domains that subsequently enter into epistemic relation. Realism privileges the world-pole; idealism privileges the mind-pole. But neither questions the architecture that produces the poles in the first place.[2]

MEOW begins by questioning the architecture.

2 The claim that the realism–idealism dispute rests on a shared and defective presupposition has been made, in different registers, by several philosophers. Rorty (1979) argued that the entire representationalist picture should be abandoned rather than reformed; Putnam, in his later work, proposed an 'internal realism' that rejected the God's-eye standpoint while retaining a commitment to objectivity; and the pragmatist tradition from James onward has treated the dispute as a pseudo-problem generated by a misleading picture of the mind's relation to the world. MEOW shares the diagnostic impulse – that the architecture is the problem – but departs from each of these predecessors in a specific respect: it proposes a replacement ontology (encounter-events as primitives) rather than a therapeutic dissolution (Rorty) or a revised epistemology (Putnam). The framework does not recommend ignoring the question; it redraws the map on which the question is located.

1.3 The Core Reorientation

The Mediated Encounter Ontology of the World proposes a simple but consequential reorientation: the most fundamental units of reality are not substances, objects, subjects, or mental states, but encounter-events.

An encounter-event is a structured relational occurrence in which three aspects co-emerge:

1. Mediation: the biological, cognitive, and linguistic–cultural organisation through which the encounter is structured.
2. Presentational structure: the way the encounter is manifested – the organised pattern through which it appears or is instantiated.
3. Constraint: the dimension of the encounter that is not freely variable – the patterned stability, resistance, or invariance that persists across variation in mediation.

These three aspects do not cause one another, nor do they stand in temporal sequence. They are structurally entailed within a single relational event. Mediation always structures some field of possibility; structured fields always exhibit constraint; and constraint can only appear within a mediational pattern. Presentational structure is their unified manifestation.

The crucial move is this: encounter-events are not relations between antecedently existing relata.[3] They are the primitive. Relata – 'mind' on one side, 'world' on the other – are abstractions retrospectively distilled from stable patterns across many such events. To see a tree, grasp a concept, or measure a physical system is not to connect two independently existing domains. It is to participate in a single organised event of encounter.

This reversal of priority shifts the explanatory burden. In substance-based metaphysics, relations derive from relata: first there are things, then they interact. In MEOW, relata derive from patterns within networks of

3 The priority of relations over relata is a commitment shared with ontic structural realism (Ladyman & Ross, 2007; French, 2014), which argues that the fundamental ontology of physics is structural rather than substantival. The parallel is real but the divergence is significant. OSR grounds its structures in the theoretical commitments of mature physics and inherits its ontological authority from scientific practice. MEOW makes no such restriction: encounter-events are not confined to physical interactions but encompass biological, cognitive, and linguistic–cultural organisation. The framework is structurally pluralistic where OSR is methodologically physicalist. Chapter 2 develops this engagement at length (see especially §2.2–2.3).

encounter: first there is structured interaction, then we abstract the 'things' that appear to participate in it. The familiar categories of 'mental' and 'physical' are therefore not ontological kinds but aspectual designations – different ways of describing the organisation of the same relational event.

1.4 Mediation as Constitutive Condition

Within MEOW, mediation is not an epistemic obstacle to be overcome. It is the structural condition under which any encounter-event arises.

Our access to reality is triply mediated. At the biological level, species-specific sensory systems, motor capacities, and neural architecture determine what kinds of encounter are possible and how they are organised. Humans encounter a world structured by trichromatic vision, binaural audition, and a particular bandwidth of electromagnetic sensitivity. Other organisms encounter differently structured worlds – the electroreceptive field of a shark, the ultraviolet-sensitive vision of a bee, the echolocative space of a bat. Each mediational architecture discloses a different encounter-organisation. None is more or less 'real' than the others; each is a different structural profile of the same underlying constraint-patterns.[4]

At the cognitive level, predictive and inferential processes organise encounter before conscious reflection begins. Perception is not passive reception but active structuring: the cognitive system generates expectations, detects deviations, and integrates sensory signals into coherent encounter-patterns. Memory, categorisation, and attentional selection further shape what is encountered and how.

At the linguistic–cultural level, inherited schemes of classification, interpretation, and valuation organise encounter at a still higher order. Colour vocabularies, spatial metaphors, temporal concepts, moral categories, institutional definitions – these are not merely labels applied after the fact

4 The idea that different organisms inhabit structurally distinct perceptual worlds has its classic formulation in Jakob von Uexküll's concept of the Umwelt (von Uexküll, 1934/2010): the species-specific environment constituted by an organism's sensory and effector capacities. Nagel's (1974) famous thought experiment – that there is something it is like to be a bat, and that this something is inaccessible from the outside – raises an adjacent question from the phenomenological side. MEOW draws on the structural insight of both: mediational architectures constitute different encounter-organisations. But it resists Nagel's conclusion that subjective character is therefore epistemically sealed. Within the framework, constraint-patterns can converge across divergent mediational architectures (see Chapter 9), and it is this convergence – not access to a shared substrate – that grounds objectivity.

but organisational structures that shape how encounters are patterned. The world encountered by a speaker of a language with twelve basic colour terms differs in its categorial organisation from the world encountered by a speaker with three, even though the constraint-patterns underlying both encounters may converge.

The philosophical significance of this triple mediation is straightforward. If every encounter is structurally mediated, then the aspiration to reach an unmediated reality behind experience is not merely difficult in practice – it is conceptually incoherent. Unmediated access would be access without organisation, which would be indistinguishable from no access at all. Mediation is the condition that makes encounter possible, not a veil to be stripped away.

This reorientation defuses a long-standing tension in metaphysics. Realism requires unmediated contact to ground mind-independent objectivity. Idealism risks reducing all structure to the activity of mind or concept. MEOW avoids the dilemma by treating mediation as ontological rather than merely epistemic: encounter-events are constituted through structured mediation, and this mediation is part of what reality is, not an obstacle to knowing it. The world does not stand behind mediation. The world appears through mediation.

1.5 Constraint as Internal Stability

Where classical idealisms often treat resistance as an inconvenience to be explained away, and realisms regard it as evidence for an external world never directly encountered, MEOW locates resistance within the organisation of the encounter-event itself.

Constraint is the dimension of an encounter that is not freely variable. It is the patterned invariance, pushback, or recalcitrance that emerges whenever a mediational structure meets a structured field of possibilities. The grain of wood under a plane, the gravitational pull on a falling body, the persistence of a perceptual illusion despite conscious knowledge of its mechanism, the impossibility of a conceptual contradiction – each is a case in which the encounter exhibits a stable pattern that cannot be willed, reframed, or interpreted out of existence.

Constraint is not imported from a noumenal realm. It is not a signal from an external world leaking through the mediation. Nor is it projected by a subjective mind. It arises from the interplay between the mediational architecture of an encounter and the structural regularities that persist across such encounters. Constraint is internal to the event's organisation: it is how patterned stability is manifested through mediation rather than behind it. Internal does not mean subjective; it means intrinsic to the relational structure of the encounter-event itself.

This framing allows MEOW to preserve empirical constraint without invoking unmediated access to a mind-independent substrate. Constraint does not signal an external object exerting influence from outside the encounter-event. It signals the structure that makes divergent encounters converge, that grounds correction and error, and that enables cross-perspectival coherence. Mediation shapes how an encounter is organised, but constraint ensures that encounters are not arbitrary.

The experience of pushback – physical, cognitive, conceptual – is therefore not evidence for something 'out there' nor merely something 'in here'. It is evidence that encounter-events have an organised structure in which not everything is up for negotiation.

1.6 Presentational Structure and Phenomenal Presentation

All encounter-events possess presentational structure: the organised pattern through which mediation and constraint are instantiated. This is the way the encounter is manifested – its structural profile, its internal organisation, the form in which it appears or could appear to an appropriate mediational system.

Presentational structure is not consciousness. A chemical reaction has presentational structure – an organised pattern of constraint and mediation at the molecular level – without any phenomenal character. A tectonic shift has presentational structure – strain patterns, displacement gradients, pressure configurations – without anything it is 'like' to be a tectonic plate. These encounters are structured, organised, and real. They are not experienced.

Phenomenal presentation is a special case of presentational structure. It arises only when mediational patterns achieve reflexive integration – when the encounter incorporates its own prior states, differentiates self from other, and integrates sensory, affective, and cognitive signals into a unified field of experience. This is characteristic of certain biological systems, though nothing in the framework restricts it to biological substrates in principle. The relevant criterion is organisational: does the system exhibit the reflexive integration that produces phenomenal character? The question is structural, not material.

This distinction is critical. It allows MEOW to claim that all encounters have structure without claiming that all encounters have experience. The framework is relational, not panpsychist. Organisation is universal; phenomenal character is not.[5] Experience is one mode of encounter-organisation among others. It is not its measure, and it is not its ground.

The original MEOW essay introduced this distinction but did not develop it at sufficient length. For the purposes of this monograph, the formulation can be stated precisely:

Presentational structure is the organised pattern of any encounter-event – conscious or otherwise.

Phenomenal presentation is the experiential character that arises when presentational structure is reflexively integrated within a mediational architecture capable of such integration.

Subsequent chapters – particularly the treatments of subjecthood (Chapter 3) and the realist defence (Chapter 9) – depend on this distinction being clearly drawn.

1.7 The Individuation of Encounter-Events

A natural question arises: how are encounter-events individuated? If they are ontologically primary, what counts as one encounter-event rather than two, or many, or a continuous field?

5 The distinction between presentational structure (which all encounters possess) and phenomenal presentation (which only some do) is what separates MEOW from panpsychism and its variants. Whiteheadian panexperientialism (Whitehead, 1929/1978) holds that all actual occasions have an experiential pole; Russellian monism (Strawson, 2006) argues that consciousness is constituted by the intrinsic nature of physical structure. Both positions treat experience as ontologically pervasive. MEOW does not. The framework claims that all encounter-events are organised – they exhibit presentational structure – but denies that organisation entails experience. The additional condition for phenomenal character is reflexive integration: the encounter must incorporate its own prior states and differentiate self from other. This is an organisational threshold, not a metaphysical primitive. The worry that MEOW's relational ontology collapses into panpsychism is addressed more fully in Appendix C (§V.F).

The answer is structural and pragmatic rather than substantial. Encounter-events are individuated by the organisation of mediation, presentational structure, and constraint that constitutes them. Two encounter-events are numerically distinct when their mediational–constraining patterns differ in structure, stability, or functional role. The criterion is not 'what is the event made of?' but 'how is the event organised?'

Because mediational architectures are layered – biological, cognitive, cultural – encounter-events can be nested, overlapping, or extended. A single perceptual scene may contain sub-events: edge-detection, colour segmentation, object-binding. Multiple observers may participate in distinct but structurally related encounters that share a constraint profile. Temporal boundaries are likewise pragmatic: an encounter-event may span milliseconds (synaptic integration), seconds (visual fixation), or hours (scientific measurement), insofar as a coherent mediational–constraining organisation persists across that duration.

This individuation scheme is intentionally pluralistic. MEOW does not impose a single metaphysical criterion for carving experience into events, because the demand for such a criterion presupposes the substance ontology the framework rejects.[6] Encounter-boundaries are fixed by the pragmatic coherence of the mediational and constraining structures relevant to explanation. An encounter may be treated as unified for one purpose – 'measuring this sample' – and decomposed for another – 'analysing error sources'. This is not relativism. It is recognition that relational organisation, not absolute metaphysical segmentation, determines what counts as an event.

Individuation is therefore neither arbitrary nor metaphysically rigid. It is grounded in the stability, coherence, and functional role of mediational–constraining organisation.

6 The claim that individuation criteria are relative to explanatory purpose has affinities with Quine's (1969) ontological relativity, which argues that what counts as an object depends on the background theory within which individuation is performed. The parallel is instructive but the motivation differs. Quine's relativity concerns reference: the question is which entities our terms pick out, and the answer is always relative to a translation manual. MEOW's pluralism concerns event-boundaries: the question is where one encounter-event ends and another begins, and the answer is relative to the mediational and constraining structures relevant to the explanatory task at hand. Quine remains within a broadly extensionalist framework; MEOW's individuation is structural and organisational. The two positions converge on the rejection of absolute metaphysical segmentation but diverge on what replaces it.

1.8 The Derivative Status of Mind and World

If encounter-events are ontologically primary, then 'mind' and 'world' are not pre-existing domains but abstractions distilled from patterns across many such events. They are not substances but stabilities – regularities that emerge when sequences of encounters display characteristic organisations.

Mind is the name we give to recurrent mediational–presentational patterns that exhibit:

- reflexive integration (encounters incorporating their own prior states)
- inference, prediction, and memory
- coherence across temporally extended encounters
- intra-encounter differentiation (self–other, figure–ground, intention–action)

These patterns cluster around a certain class of encounter-systems, but nothing metaphysical hinges on their biological instantiation. 'Mind' identifies a stable organisational profile, not a substance that stands behind the profile. In this sense, mind is emergent: the result of patterned regularities across encounter-networks, not a primitive ontological kind.

World, analogously, names the constraint-dominated regularities that persist across variations in mediation:

- cross-perspectival invariances
- stable causal–temporal sequences
- patterns that resist variation across biological, cultural, or instrumental differences
- constraint structures exhibited in physical, chemical, and mathematical regularities

To call something 'worldly' is not to posit a substance behind experience. It is to say: this pattern of constraint recurs across encounter-events irrespective of the mediational differences between them. This is the operative meaning of objectivity within MEOW: stability across mediation, not access to an unmediated substrate.

Both abstractions are useful. Both are, in a sense, indispensable – human thought and social life would be unmanageable without them. But they track recurring profiles of encounter, not independently existing domains.

The monograph's subsequent chapters develop the consequences of this derivative status across epistemology, logic, perception, language, and social ontology. For the present, the point to retain is structural: 'mind' and 'world' are not where the analysis begins. They are where it arrives, after encounter has done its work.

1.9 Formal Summary of Core Commitments

The central commitments of MEOW can be stated in condensed form. These are not additional claims but a compression of the framework developed above.

Thesis 1. Ontological Primacy of Encounter-Events. All that exists exists through encounter-events: structured relational occurrences in which mediation, presentational structure, and constraint co-emerge. Relata are abstractions over patterns within such events, not ontological primitives.

Thesis 2. Mediation as Constitutive Structure. Every encounter-event is organised through mediation – biological, cognitive, and linguistic–cultural. Mediation is not a veil between mind and world but the structural condition under which encounter becomes possible. No access occurs outside mediation.

Thesis 3. Constraint as Intrinsic Stability. Constraint arises within encounter-events themselves. It is the dimension of stability, resistance, or invariance that is not freely alterable by variation in mediation. Constraint is not projection and not a noumenal intrusion.

Thesis 4. Co-Emergence of Mediation, Presentational Structure, and Constraint. Mediation, presentational structure, and constraint are not sequential components but aspectual descriptions of a single relational organisation. No aspect is metaphysically prior; all arise jointly in the event's organisation.

Thesis 5. Derivative Status of 'Mind' and 'World'. 'Mind' and 'world' designate stable patterns abstracted from regularities across encounter-events rather than independently existing substances. Neither term denotes a domain external to encounter.

Thesis 6. Rejection of the Mental–Physical Dichotomy. 'Mental' and 'physical' are aspectual predicates applied to different emphases within encounter-event organisation. They identify classificatory stances, not ontological kinds.

Thesis 7. Relational Closure. All experiential, epistemic, and causal processes occur within networks of encounter-events. There is no need – and no conceptual room – for positing extra-encounter substances or forces.

Thesis 8. Causal Dependence as Stable Encounter Sequencing. Causation is not substance-level production but the stability of constraint transmission across sequences of encounter-events. An encounter E_1 is causally relevant to E_2 when the constraint-profile of E_1 reliably shapes the possible configurations of E_2 across mediational variations.[7]

1.10 What This Chapter Has Established

This chapter has presented the foundational commitments of the Mediated Encounter Ontology of the World in a form intended to be self-sufficient for the monograph that follows.

The framework begins from the recognition that the realism–idealism impasse is sustained by a shared presupposition – the independent existence of mind and world as prior domains – and proposes an alternative: encounter-events as ontologically primary, with mediation, presentational structure, and constraint as their co-emergent aspects.

'Mind' and 'world' are derivative abstractions. Mediation is constitutive, not obstructive. Constraint is internal to encounter, not imported from a noumenal exterior. Phenomenal presentation is a special case of presentational structure, arising under reflexive integration, not a universal feature of all encounter.

---

7 The deflationary treatment of causation developed here is stated in compressed form. The full account, including its relationship to the Humean regularity tradition, engagement with the powers metaphysics of Mumford and Anjum (2011), and the framework's handling of probabilistic and counterfactual causation, is presented in §2.9. The reader who finds Thesis 8 insufficiently motivated at this stage is invited to treat it as a promissory note redeemed in the following chapter.

Much remains to be developed. The ontological commitments introduced here require a fuller treatment of what it means for reality to be a subjectless field of interactional processes – and in particular, a defence against the objection that relations without relata are unintelligible. That is the task of Chapter 2.

The architecture of encounter – subjecthood, logic, epistemology, perception, language – requires its own sustained treatment in Chapters 3 through 7. The realist defence of the framework, the demonstration of its explanatory purchase, and an honest accounting of its limits occupy the monograph's final movement.

What this chapter offers is the ground. The subsequent chapters build on it.

2

Ontology

Interaction, Constraint, and the Rejection of Substance

The many become one, and are increased by one.

— Alfred North Whitehead, *Process and Reality*

Life's not a static object, after all. It's a process.

— Ursula K Le Guin, *The Lathe of Heaven*

2.1 Introduction: The Question Behind the Framework

Chapter 1 established that encounter-events are ontologically primary, and that 'mind' and 'world' are derivative abstractions over stable patterns within networks of such events. The immediate question is: what kind of ontology does this commitment entail?

The question is more pointed than it first appears. MEOW rejects substance ontology – the view that reality consists of independently existing things that bear properties and enter into relations. It rejects the mind–world bifurcation that substance thinking generates. It proposes instead that reality consists of structured interactional processes, and that what we call objects, subjects, and categories are scale-dependent stabilisations within a subjectless field of relational constraint.

But that proposal invites scrutiny. If there are no substances, what is doing the interacting? If relations are primitive, what are the relations relations of? If the 'field' is subjectless, what kind of thing is it? These are not hostile questions. They are the questions any serious ontology must answer.

This chapter answers them. It develops the ontological commitments of MEOW in full, defends the coherence of relational primacy against the most pressing objection in the literature, and clarifies the status of the framework's central terms – particularly the phrase 'subjectless field of relational constraint', which will do significant work throughout the monograph.

A preliminary clarification is therefore warranted before the argument proceeds. The phrase 'subjectless field of relational constraint' is not a physics hypothesis. It does not posit a literal field in the sense of quantum field theory or classical electromagnetism. Nor is it a metaphor – a loose gesture toward something ineffable that the framework declines to specify. It is an ontological commitment: reality, at its most basic level, is structured interaction, and this interaction does not require pre-existing substances to sustain it. The word 'field' is used to indicate that relational constraint is not localised in discrete objects but distributed across encounter-organisation.[1] The word 'subjectless'

1 The choice of 'field' invites comparison with quantum field theory, which the main text explicitly disclaims. It also invites comparison with Gestalt psychology, which used 'field' to describe the structured totality of perceptual experience (Köhler, 1947; Lewin, 1951). The parallel is closer than it might appear: Gestalt field theory insisted that perceptual organisation is not assembled from pre-existing elements but arises as a structured whole – a claim structurally homologous to MEOW's treatment of encounter-events. The framework does not draw on Gestalt theory directly, but the convergence is worth noting for readers

indicates that this organisation does not depend on observers, experiencers, or conscious systems for its existence – though it requires encounter for its disclosure. The distinction between existence and disclosure is central to MEOW and will be developed throughout this chapter. Stated precisely: existence is invariant constraint-pattern; disclosure is its articulation within mediation. The constraint-pattern does not depend on encounter for its persistence. It depends on encounter for its manifestation.

2.2 What Exists: Structured Interactional Processes

The ontological primitive of MEOW is not a thing but a pattern of interaction.

This is less exotic than it sounds. Modern fundamental physics does not describe the world as a collection of independently existing substances in the classical sense. In quantum field theory, the basic entities are fields and their excitations, and what are commonly referred to as 'particles' are understood as stable patterns – quanta – within these fields. The 'particle' is the name we give to a recurring configuration, not a miniature object. Interaction, rather than substance, plays the organising role in the theoretical description. As a consequence, many contemporary philosophers of physics have argued that the ontological commitments suggested by modern physics are structural or relational rather than substantival (Ladyman & Ross, 2007; French, 2014).

This monograph does not attempt to derive its ontology from physics. It merely observes that the relational orientation developed here is broadly consistent with the direction in which contemporary scientific description has already moved. MEOW generalises this insight beyond physics – not because it borrows from physics, but because the same structural observation applies wherever we look. What we encounter as stable, object-like entities are, on closer inspection, dense and slow-changing constraint clusters within broader fields of interaction. A boulder is a stable constraint cluster at the human scale of encounter. At the particle scale it dissolves into probabilistic configurations. At the geological scale it flows. At the neutrino scale it is barely there at all. The boulder is real – one can stub a toe on it – but its reality is that of a scale-dependent stabilisation, not a substance that exists independently of any encounter-context.

familiar with that tradition. The operative sense of 'field' in MEOW is distributive: constraint is not localised at a point but patterned across a relational organisation. This is a topological metaphor, not a physical hypothesis.

This does not mean that objects are illusions. It means that their ontological status is that of stabilities rather than substances. They are patterns that persist within certain encounter-regimes, not things that exist prior to and independent of all encounter. Substance ontology treats persistence as primitive and change as derivative. Beyond MEOW reverses this priority: stability is slow interaction, and identity is stabilised duration.[2] The distinction matters because it determines how we understand objectivity, persistence, and identity – questions developed in later chapters. For now, the point is architectural: what exists, at the most basic level, is not stuff but structured interaction. Objects are what happens when structured interaction stabilises within a given scale of encounter.

2.3 The Invariant Anchor: Relational Constraint

If objects are stabilisations rather than substances, and if the field of interaction is subjectless – not dependent on observers for its existence – then what grounds the framework's realism? What prevents MEOW from collapsing into a sophisticated form of constructivism in which everything is relative to encounter and nothing is genuinely invariant?

The answer is relational constraint.

What persists across encounters is not substance but patterned invariance – the structural regularities that remain stable across variations in mediation. Gravitational constraint persists whether the encountering system is a human with a mass spectrometer, a beetle on a branch, or a satellite in orbit. Decay constants remain constant regardless of the conceptual framework used to model them. The Müller-Lyer illusion persists despite

2 The formulation 'stability is slow interaction, and identity is stabilised duration' is the framework's compression of a structural insight with a long pedigree. Heraclitus' dictum that one cannot step into the same river twice expresses the pre-Socratic version of the claim that persistence is a mode of process rather than a substrate beneath it. Whitehead's process metaphysics (1929/1978) systematised the insight: actual occasions are the primitive, and enduring objects are 'societies' of occasions exhibiting serial order. Bergson's critique of spatialised thinking, developed in *Time and Free Will* (1889/1910), argues that the intellect's tendency to freeze process into static categories distorts the nature of duration. MEOW inherits the structural commitment from all three traditions – that process is prior to substance – but differs from each. It does not share Heraclitus' monism of flux, Whitehead's panexperientialist metaphysics, or Bergson's intuitionist epistemology (on Bergson, see §7.5). What it shares is the reversal of explanatory priority: stabilities are achievements within a dynamic field, not the field's foundation.

conscious knowledge of its mechanism. These are not projections. They are constraint-patterns that recur across encounters organised by radically different mediational architectures.

The principle can be stated precisely: filtering does not imply fabrication. Mediation shapes how an encounter is organised – what is foregrounded, how it is categorised, which aspects are salient – but it does not manufacture the constraint-patterns that persist across mediational variation. A trichromatic visual system and a tetrachromatic visual system will encounter colour differently, but the spectral regularities that constrain both encounters are not produced by either system. They are what makes convergence across the two systems possible.

This is MEOW's realist anchor. Not correspondence to an unmediated reality – that aspiration is incoherent, as Chapter 1 argued – but convergent stability across distinct mediational architectures. When differently organised encounter-systems converge on the same constraint-patterns, that convergence is evidence of invariant structure rather than coincidence of projection. The full defence of this anchor, including its relationship to scientific realism and constructivism, is developed in Chapter 9. For the present, it is sufficient to establish the principle: relational constraint is the invariant core of the framework, and it is what prevents MEOW's rejection of substance from entailing a rejection of realism.

2.4 The Ontology/Model Distinction

A further clarification is needed – one that will be less obvious to sympathetic readers than to hostile ones, but that matters in both cases.

MEOW's ontological claims are not claims about models. When the framework asserts that reality consists of structured interactional processes, it is not offering a model of reality that might be replaced by a better model tomorrow. It is making a commitment about what is ontologically basic: interaction, not substance. This is a claim about the architecture of what exists, not about the best way to represent it for predictive purposes.

The distinction matters because MEOW operates in a space where several adjacent positions can be confused with it.

Ontic structural realism (Ladyman and Ross) also rejects substance ontology in favour of structural relations. But OSR grounds its structures in the commitments of fundamental physics, treating the mathematical structures of our best physical theories as the ontological bedrock. MEOW does not. Encounter-organisation arises in perception, chemistry, geology, computation, and social interaction – not only in the domain of fundamental physics. MEOW is structurally pluralistic where OSR is methodologically physicalist. The two frameworks share an anti-substantivist orientation, but they diverge on scope and on the privileging of physics as the court of final ontological appeal. MEOW shares structural realism's rejection of substance ontology but grounds relational structure in encounter-organisation rather than in the mathematical structure of physical theory. The distinction is not terminological. It determines the framework's range.

Scientific realism in its standard forms maintains that mature scientific theories describe (approximately) mind-independent entities and structures. MEOW is compatible with the practice of scientific realism – it has no quarrel with the utility of positing electrons, genes, or tectonic plates for explanatory purposes – but it reframes the ontological status of such posits. They are stable constraint-patterns within encounter-networks, not substances existing independently of all encounter. Science tracks constraint-structure; it does not require the further claim that what it tracks exists in a mind-independent realm behind encounter.

Constructivism holds that the categories through which we understand reality are, to a significant degree, constructed by the cognitive or cultural systems that deploy them. MEOW agrees that mediation shapes encounter-organisation. But it insists – and this is the point at which it parts company with constructivism – that constraint is not reducible to construction. Mediation determines the form of an encounter; constraint determines what is stable within it. Constructivism overstates the sovereignty of the mediational system. MEOW balances mediation with constraint.

Kantian noumenalism posits an unknowable reality behind appearances. MEOW rejects this.[3] There is no hidden realm of things-in-themselves

3 The rejection of noumenalism is a commitment the framework shares with several traditions that otherwise diverge from it. Hegel objected that the thing-in-itself is an abstraction produced by the very understanding it was supposed to limit. The pragmatists – James, Dewey, Rorty – argued that the noumenal posit does no explanatory work. Sellars' (1956/1997) critique of the 'myth of the given' undermines the perceptual foundations on which noumenal inference typically rests. MEOW's rejection is not borrowed from any of these but converges with all of them: the noumenal–phenomenal distinction is, within

behind encounter-events. Constraint is internal to encounter-organisation, not a trace of something transcendent. The distinction between what is encountered and what is projected is preserved – but it is drawn within encounter, not between encounter and an inaccessible exterior.

What MEOW claims is therefore this: reality is structured interaction, encounter-events are the primitive locus of that interaction, and relational constraint is what makes the interaction invariant rather than arbitrary. This is not a model. It is not a metaphor. It is not a repackaging of an existing position. It is an ontological commitment about what there is, stated as precisely as the framework permits.

2.5 The No-Relata Objection

The most technically pressing objection MEOW faces is the no-relata problem, developed most forcefully by Anjan Chakravartty against ontic structural realism[4] and applicable, with appropriate modification, to any framework that treats relations as ontologically primitive.

The objection runs as follows. Relations, as ordinarily understood, require relata – the things that stand in the relation. If I say 'A is taller than B', the relation 'taller than' presupposes the existence of A and B as independently identifiable entities. To claim that relations are primitive and relata are derived is, on this view, to claim that there are relations without anything being related – which appears to be either incoherent or an unacknowledged reintroduction of substance under another name.

The objection deserves a serious response. It is not a misunderstanding. It identifies a genuine structural demand: if relations are primitive, the framework must explain what it is that the relations are relations of, or it must show that the demand itself is misconceived.

the framework, a post hoc artefact of substance thinking rather than a structural feature of reality. The full treatment of this claim is in Chapter 10 (§10.7).

4 Chakravartty's (2007) objection targets OSR specifically, arguing that structures require bearers and that purely structural ontology is either incoherent or tacitly reintroduces individuals. The objection has generated a substantial literature: Esfeld (2004) defends a 'moderate' structural realism that retains relata as individuated by their relational properties; French (2014) responds with a reconceptualisation of objecthood as structural. MEOW's response takes a different path from either: it does not retain relata as thinly individuated (Esfeld) or reconceptualise objects as structural placeholders (French), but treats the demand for relata as a diagnostic symptom of the substance ontology the framework rejects. The whirlpool analogy in the main text is intended to make this structural, not merely rhetorical.

MEOW takes the second path. The demand for relata behind relations is not a neutral methodological requirement. It is a substantival commitment disguised as a logical one. It assumes that the basic form of reality is thing-bearing-property or thing-standing-in-relation, and then asks the relational ontologist to identify the things. But this is precisely the architecture MEOW rejects. The demand presupposes what is at issue.

Consider an analogy.[5] A whirlpool is a real, identifiable, causally efficacious pattern.

MEOW applies the same logic at the ontological level. Encounter-events are structured relational occurrences. What recurs across encounters is patterned constraint – invariance, resistance, stability. The 'relata' that appear within encounters – the objects, the subjects, the properties – are stable configurations within these patterns, not pre-existing entities that the patterns connect. To demand relata behind the relations is like demanding the whirlpool's skeleton. The demand reflects a category expectation shaped by substance thinking, not a genuine structural requirement of the ontology.

This response does not dodge the objection. It diagnoses it. The no-relata problem arises because the question 'what are the relations relations of?' is formulated within the very framework – substance ontology – that relational ontology rejects. It is not that the question is unanswerable. It is that the question is wrongly framed. Within MEOW, the correct formulation is not 'what are the relations between?' but 'what stable patterns emerge within relational organisation?' The answer is: objects, subjects, properties, and all the other furniture that substance ontology treats as primitive. They are real. They are explanatorily indispensable. But they are derived, not basic.

The firmness of this response should not be mistaken for evasion.

5 Concrete illustrations inevitably involve entities that appear to function as relata – water, rocks, magnetic fields. This reflects the grammatical bias analysed in Chapter 7: natural language strongly favours noun-based encoding, and relata-free description is articulatorily resistant in any Indo-European language. The examples are pedagogical scaffolding rather than ontological commitments. The formal commitment is to relational primacy as stated in A1–A2 of Appendix A. in a body of water. One can point to it, measure it, drown in it. But it is not an object in the substance-ontological sense. It is a stable pattern of motion – a constraint-configuration within a dynamic medium. To ask 'what is the whirlpool made of?' is to ask a question that the phenomenon does not reward. The whirlpool is water in motion; the motion is what makes it a whirlpool; remove the motion and you have water but no whirlpool. The 'substance' of the whirlpool is the pattern, not some additional thing that the pattern is a pattern of.

MEOW does not deny that relata exist. It denies that they are ontologically prior. Tables, chairs, persons, and electrons are all real – they are stable constraint-clusters within encounter-networks. What they are not is the starting point of ontology. The starting point is the encounter-structure from which they are abstracted.

A sceptic might press further: if relata are abstracted from relational patterns, what prevents the abstraction from being arbitrary? What ensures that the 'objects' distilled from encounter-networks are not merely convenient fictions? The answer is constraint. Not all abstractions are equally stable. A boulder persists across a wide range of mediational variations; a conspiracy theory does not. The constraint-patterns that underwrite the abstraction determine its robustness. The invariant anchor – convergent stability across distinct mediational architectures – is what separates well-grounded abstractions from projections. This is MEOW's structural replacement for the substance ontologist's appeal to mind-independence: not 'the object exists independently of all encounter' but 'the constraint-pattern recurs across encounters organised by different mediational systems'.

2.6 Scale-Dependence and the Ontology of Objects

If objects are scale-dependent stabilisations, the natural question is: stabilisations of what, at what scale, and for whom?

The 'of what' is now answered: stabilisations of relational constraint-patterns within encounter-networks. A boulder is a dense, slow-changing constraint cluster at the human scale of encounter – at the scale determined by our morphology, our energetic interactions with the environment, and our perceptual bandwidth. It is real at that scale. Its reality is not diminished by the observation that at other scales – particle physics, geological deep time, neutrino interaction – it dissolves, flows, or vanishes. Scale-dependence is not a defect in the object's reality. It is a feature of what it means for reality to be structured interaction rather than substance.

The 'at what scale' is determined by the mediational architecture of the encountering system. Human-scale objects – trees, rivers, tools, buildings – are stabilisations relative to organisms of our size, energetic engagement, and perceptual organisation. Other organisms encounter different stabilisations. A bacterium does not encounter a boulder; it encounters surface chemistry.

An eagle does not encounter a landscape the way a mole does. Each encounter-system abstracts its objects from the constraint-patterns available at its mediational scale.

This is not perspectivism in the relativistic sense. The constraint-patterns are not created by the encountering system. They are disclosed by it – structured through mediation, but not manufactured by mediation. The boulder's constraint-pattern does not depend on human observation for its existence. But its individuation as 'a boulder' – as a discrete, bounded, persistent object – does depend on a mediational architecture operating at the appropriate scale. Existence and individuation are not the same thing. The constraint-pattern exists independently of encounter; its articulation as a particular object is encounter-relative.[6]

This distinction – between existence of constraint and articulation of objects – is central to MEOW's realism. It allows the framework to maintain that something is genuinely there (invariant relational constraint) without requiring that the way it is carved into objects is uniquely determined by reality itself. Different mediational architectures carve differently. What they carve from – the constraint-field – is not up for negotiation.

2.7 Why Substance Thinking Persists: A Preliminary Note on Ontological Grammar

If substance ontology generates the problems diagnosed in Chapter 1 – the mind–world bifurcation, the realism–idealism impasse, the no-relata demand itself – why does it persist? Why is it so difficult to think past?

Part of the answer is institutional. Substance ontology is embedded in the curricula, the journals, and the conceptual reflexes of professional philos-

6 The distinction between existence and individuation bears a structural resemblance to Bhaskar's (1975/2008) stratified ontology in critical realism, which distinguishes between the real (mechanisms and structures), the actual (events), and the empirical (experienced events). Bhaskar's 'real' domain – generative mechanisms that exist whether or not they are activated or observed – parallels MEOW's constraint-patterns, which persist independently of their disclosure in encounter. The divergence is that Bhaskar retains a broadly substance-ontological framework: his generative mechanisms are properties of structured things. MEOW dissolves the thing-bearing-mechanism picture and locates the generative work in relational constraint itself. The parallel is instructive precisely because it shows how far a realist ontology can travel without abandoning substance – and where it must stop.

ophy. Careers are built on refining positions that presuppose it. This is a sociological observation, not an argument, but it explains some of the resistance.

A deeper part of the answer, however, is grammatical. The languages in which philosophy has historically been conducted – and in which this monograph is written – impose substance-ontological commitments through their basic syntactic structure. Indo-European languages are built on subject-predicate form: a noun (substance) bears a verb (action) or an adjective (property).[7] 'The rock falls.' 'The mind perceives.' 'The electron orbits.' Each sentence presupposes an entity that exists prior to and independent of what it does. The grammar makes substance ontology feel inevitable before any philosophical reflection has begun.

This is not merely a claim about convenience or habit. It is a claim about the structural conditions of thought as conducted in natural language. When a philosopher writes 'the mind perceives the world', the subject-predicate structure of the sentence has already installed two substances – mind and world – and a relation between them. The question then becomes how the relation works, which is the realism–idealism dispute in miniature. But the architecture was supplied by the grammar, not by the phenomena. A language that did not organise experience into noun-subjects and verb-predicates would not generate the same metaphysical landscape.

The full development of this argument – including its implications for logic, institutional reasoning, and the limits of philosophical language – is the task of Chapter 7, which treats language as a finite-bandwidth synchronisation protocol and examines the ontological commitments smuggled in by its grammatical structure. For the present, the point is diagnostic: the persistence of substance ontology is not merely a philosophical failure of imagination. It is partly an artefact of the grammatical medium in which phil-

7 The claim that grammatical structure shapes ontological intuition is developed at length in Chapter 7 (§7.3–7.6) under the heading 'ontological grammar'. It should be noted here that the claim is not equivalent to linguistic determinism in the strong Whorfian sense. The framework does not hold that speakers of Indo-European languages cannot think in relational or processual terms – manifestly they can, as the existence of process philosophy, Buddhist metaphysics, and this monograph itself demonstrates. The claim is about the path of least resistance: subject-predicate grammar makes substance-ontological formulations fluent and relational formulations cumbersome. The asymmetry in articulatory cost biases philosophical reflection toward substance categories without requiring that it be confined to them. This is an effectiveness claim, not a determinism claim. See §7.4 for the full distinction.

osophical imagination operates. Recognising this does not refute substance ontology – arguments, not etymologies, do that work – but it explains why the refutation meets resistance that seems disproportionate to the strength of the position being defended.

2.8 Filtering Does Not Imply Fabrication

A principle introduced in Section 2.3 deserves its own brief treatment, because it will recur throughout the monograph and it is the hinge on which MEOW's realism turns.

Mediation shapes encounter. This is a commitment of the framework, not a concession to its critics. Every encounter is organised through biological, cognitive, and linguistic–cultural mediation. The world as encountered is always already structured by the conditions of encountering.

But shaping is not manufacturing. A filter alters what passes through it; it does not create the material being filtered. A trichromatic visual system shapes colour-encounter differently from a tetrachromatic one. But the spectral regularities that both systems respond to are not produced by either. A scientific instrument shapes the data it collects – amplifying some signals, attenuating others, introducing systematic artefacts. But the constraint-patterns the instrument detects are not invented by the detection process.

The principle – filtering does not imply fabrication – is MEOW's answer to two opposing anxieties.

The realist worries that if all encounter is mediated, nothing genuinely independent is ever reached. MEOW replies: what is reached is not independent of mediation (nothing could be), but the constraint-patterns that recur across different mediations are not artefacts of any particular one. Independence is replaced by convergent invariance. That is enough to ground objectivity.

The constructivist worries that if constraint is real, mediation is reduced to a transparent window and the constructive contribution of cognitive and cultural organisation is denied. MEOW replies: mediation is constitutive – it genuinely shapes encounter-organisation. But it shapes it within constraint, not against an empty background. The constructive contribution is real. It is just not sovereign.

This principle is not a compromise. It is a structural feature of the ontology. Mediation and constraint are co-emergent aspects of encounter-events. Neither is reducible to the other. Neither is epiphenomenal. The realist and the constructivist each see half the picture. MEOW insists on the whole of it.

2.9 Causation as Constraint Transmission

If substances are rejected, what becomes of causation? Substance ontology understands causation as the production of effects by independently existing agents: billiard ball A strikes billiard ball B and causes B to move. If there are no billiard balls at the fundamental level – only relational constraint-patterns – how is causal structure preserved?

MEOW reframes causation as the stable sequencing of constraint-patterns across encounter-events.[8] An encounter-event E_1 is causally relevant to E_2.

Causal asymmetry – the fact that causes precede effects and not the reverse – arises from the organisation of mediation itself. Encounter-events unfold within temporally oriented networks: sensory uptake precedes integrative processing; experimental preparation precedes measurement; environmental change precedes adaptive response. These asymmetries belong to the architecture of encounter, not to substances exerting influence from without. What persists across contexts is not a metaphysical arrow but a directional stability in how encounter-events transition.

Probabilistic causation is accommodated without difficulty. Where transitions from E_1 to E_2 show statistical rather than deterministic regu-

8 This deflationary treatment extends a sceptical tradition originating with Hume, who argued that the 'necessary connexion' we attribute to causal relations is a psychological expectation projected onto constant conjunction rather than a feature observed in experience. Hume's suspicion pushed subsequent debate toward regularity, counterfactual, and process accounts of causation – each attempting to preserve causal structure without invoking occult powers. The present formulation sits in that lineage: what we call causation is the stable propagation of constraint-patterns across encounter-events. Nothing metaphysically thicker is required. Contemporary physics quietly leans in the same direction – field interactions, conservation laws, and dynamical constraints specify how systems evolve without invoking causal essences. They describe what transitions are allowed and how states constrain later states, which is closer to the framework's vocabulary than the traditional metaphysical picture of hidden causal powers. when the constraint-profile of E_1 reliably shapes, limits, or organises the possible configurations of E_2 across variations in mediation. This is not a claim about hidden powers or noumenal mechanisms. It is a claim about the regularity of transition-patterns within encounter-networks.

larity, causal dependence is expressed as probabilistic constraint: mediation structures the range of possible transitions, and constraint structures their relative likelihoods. Omissions and preventions register as counterfactual disruptions in expected encounter-sequences, rather than as the absence of substance-interaction.

This account of causation is deliberately deflationary. It does not seek to discover hidden causal powers behind the regularities.[9] It articulates the structure that underlies causal inference

2.10 Temporal Organisation

One further structural feature of encounter-events requires treatment before this chapter closes: temporal organisation.

Encounter-events do not unfold within a pre-existing temporal container. Rather, temporal structure emerges within the organisation of mediation itself.[10] Relations such as simultaneity, succession, and duration are features of how encounters are patterned, not properties imposed from outside the encounter.

Different mediational structures organise temporal sequence differently. Perceptual time includes the specious present and the binding of successive inputs into apparent continuity. Biological time expresses developmental

9 Dispositional and powers-based accounts of causation (e.g., Mumford & Anjum, 2011) offer an alternative approach grounded in causal capacities rather than regularities. The framework's deflationary treatment is a deliberate architectural commitment, not a claim that such accounts are incoherent. – stable constraint-transitions, directional organisation of mediation, probabilistic constraint, counterfactual sensitivity – and claims that this structure is all that causation is. Scientific practice relies on exactly these features. MEOW makes them explicit without adding metaphysical surplus.

10 The claim that time is internal to encounter-organisation rather than an external container places MEOW in the neighbourhood of relational theories of time, which have a long history in both physics and metaphysics. Leibniz argued against Newton's absolute time, maintaining that temporal relations are constituted by the ordering of events rather than by a pre-existing temporal substance. In contemporary philosophy of physics, Rovelli's (1996) relational quantum mechanics eliminates time as a fundamental variable in favour of correlations between physical systems. MEOW's treatment is less technically specific than either – it does not commit to a particular physical formalism – but it shares the structural orientation: temporal organisation is a feature of how encounters are patterned, not a stage on which they are performed. The framework's deliberate restraint on this topic ('minimal by design', as the main text notes) reflects the fact that a full encounter-theoretic treatment of temporality would require engagement with the physics of time that lies beyond this monograph's scope.

rhythms, circadian cycles, and metabolic pacing. Physical time introduces precision via coordinate systems, decay rates, and dynamical equations. These temporal modalities are not competing descriptions of a single temporal substrate. They are distinct modes of temporal organisation within encounter-networks.

Past-tense claims are retrodictive structures within present encounters, grounded in the stability of constraint-patterns rather than access to a bygone temporal segment. The statement 'dinosaurs existed sixty-five million years ago' is warranted not by stepping outside encounter to inspect the past but by the convergent constraint-patterns of present encounters: fossil morphology, isotopic decay curves, geological stratification, cross-confirming measurement systems. These present constraints support retrodictive inference – not to mind-independent substances located in the past, but to temporal patterns consistent with the constraint-profiles currently observed.

This treatment of temporality is minimal by design. A fuller account of temporal organisation – including the relationship between encounter-time and physical time, the structure of memory, and the temporal dimension of social institutions – lies beyond the scope of this chapter. What matters here is the structural point: time, like objects and subjects, is internal to encounter-organisation rather than a container within which encounters occur.

2.11 What This Chapter Has Established

This chapter has developed the ontological commitments of MEOW beyond the foundation laid in Chapter 1.

Reality is a subjectless field of structured interactional processes. This is an ontological commitment, not a physics claim, a metaphor, or a disguised version of ontic structural realism. Objects are scale-dependent stabilisations of relational constraint-patterns. Relational constraint – patterned invariance across mediational variation – is the invariant anchor of the framework. The principle that filtering does not imply fabrication preserves realism without requiring unmediated access. Causation is stable constraint-transmission across encounter-sequences. Temporal organisation is internal to encounter rather than externally imposed.

The no-relata objection – the demand for substances behind relations – has been addressed directly. The demand presupposes the substance ontology MEOW rejects. Relations are primitive; relata are stable abstractions within relational organisation. This is coherent, not evasive, and the constraint-patterns that underwrite the abstraction determine its robustness.

The chapter has also introduced, in preliminary form, the observation that the persistence of substance thinking is partly a grammatical phenomenon – a consequence of the ontological commitments embedded in Indo-European subject-predicate structure. This claim is developed fully in Chapter 7.

What has not yet been addressed is the question of subjecthood. If objects are stabilisations within encounter-networks, what is the status of the encountering system itself? How does the interactional field become differentiated into the perspective of a subject – a locus from which encounter is organised, affordances are disclosed, and the world takes on salience? If relational constraint is primary, then subjecthood must be explained as differentiation within that field – not as an ontological exception to it. That is the task of Chapter 3.

ARCHITECTURE

II

3

Subjecthood

Modal Differentiation Within the Field

I is another.
Je est un autre.

— Arthur Rimbaud, letter to Paul Demeny

Do I contradict myself?
Very well then I contradict myself,
(I am large, I contain multitudes.)

— Walt Whitman, *Song of Myself*

3.1 Introduction: The Missing Agent

The preceding chapters have developed an ontology of structured interaction and established its consequences for language and institutions. Encounter-events have been characterised. Constraint has been grounded. The synchronisation protocol has been analysed and its limits identified. Institutions have been shown to be second-order constraint systems built from that protocol.

But a question has been quietly accumulating. Throughout this architecture, 'organisms' and 'agents' have appeared as participants in encounter-events, users of the synchronisation protocol, and subjects of institutional constraint. They have not yet been explained. The framework owes an account of what these entities are – and, more precisely, of what it means to be a subject within an ontology that rejects substance.

The question is unavoidable. If objects are scale-dependent stabilisations of relational constraint, what is the status of the system that encounters them? If 'mind' is a derivative abstraction over encounter-patterns, what is doing the encountering? If language synchronises stabilisations across agents, what kind of thing is an agent?

The substance-ontological answer is familiar: a subject is a thing – a soul, a self, a mind-substance, a Cartesian ego – that exists independently of its encounters and subsequently enters into relations with the world. MEOW rejects this answer. But rejecting it creates an obligation. If the subject is not a substance, what is it? And if it is not a substance, is it real?

This chapter argues that subjecthood is real but relational. It is not a thing inside the architecture. It is a differentiation within the architecture – a specific organisational profile that emerges when certain process-clusters achieve sufficient complexity to generate perspective, persistence, and agency within the constraint-field. The subject does not observe the field from outside. It is a dynamic patterning within the field that creates the conditions for encounter to become oriented – to have a direction, a salience structure, a perspective from which some things matter and others do not.

3.2 Modal Differentiation

The concept at the centre of this chapter is modal differentiation. It requires careful definition because it is doing work that substance ontology distributes across several distinct concepts – self, consciousness, agency, perspective – which MEOW treats as aspects of a single organisational profile.

Modal differentiation is the point at which the interactional field becomes differentiated into constraint and affordance relative to a self-maintaining, constraint-responsive process cluster. A terminological ladder will help prevent slippage in what follows. A process-cluster is any stabilised pattern within the field. A self-maintaining cluster is one that actively preserves its own organisation against perturbation. A modally differentiated cluster is a self-maintaining cluster whose activity generates salience and affordance asymmetry within the field. A subject is a modally differentiated cluster that exhibits reflexive integration – recursive modelling of its own persistence conditions as an input to action selection across counterfactual affordance space.[1] Subjecthood, in other words, requires not merely self-maintenance but the organisation of the field as a space of possible moves relative to the maintenance problem, including anticipated futures.

Before differentiation, there is structured interaction – relational constraint patterns, encounter-events, presentational structure – but no perspective. After differentiation, the same field acquires directionality: some interactions become salient, some pathways become viable, some constraint-patterns become relevant and others recede. The field does not change. What changes is that a locus has emerged within it from which the field is organised asymmetrically.

1 The concept of modal differentiation draws on, but is not identical to, the autopoietic tradition initiated by Maturana and Varela (1980) and extended by Thompson (2007). Autopoiesis identifies self-production as the defining characteristic of living systems: an autopoietic system continuously generates and regenerates the components that constitute it. MEOW's 'self-maintenance' is structurally adjacent but wider in scope. It does not require metabolic self-production in the strict autopoietic sense; what it requires is active preservation of organisational coherence against perturbation. This permits substrate-neutral application. A system that maintained its own organisational coherence through non-metabolic means – were such a system to exist – would qualify as self-maintaining within the framework without satisfying the autopoietic criterion. The distinction is deliberate: MEOW needs an organisational concept that is broader than biology without being so broad as to include thermostats.

An analogy may help, provided it is not pressed too far. A magnetic field exists uniformly throughout a region of space. Place a compass within it and the field acquires local directionality – not because the field has changed but because a responsive system has been introduced that differentiates the field relative to its own structure. The compass does not create the field. It does not observe the field from outside. It is a physical system within the field whose internal organisation makes the field's structure salient in a particular direction. The compass analogy captures orientation, not concern. Concern enters only with self-maintenance under threat – when the system bears the cost of its own persistence and the field therefore acquires valence, not merely direction.[2]

Modal differentiation works similarly but at the level of encounter-organisation. A self-maintaining process cluster – paradigmatically a living organism, though the formulation is substrate-neutral – generates a local differentiation of the constraint-field into what is navigable and what is obstructive, what is nourishing and what is toxic, what affords action and what resists it. This differentiation is not added to the field from outside. It is an organisational feature that arises when a sufficiently complex, self-maintaining system operates within the field. The subject is not an observer of the architecture. It is a node of differential organisation within it.

The term 'modal' is chosen deliberately. The differentiation is not spatial (the subject is not located at a geometric point) nor temporal (it is not a moment in a sequence). It is modal: it concerns the way the field is organised relative to the process-cluster's own persistence conditions. Modal here is not logical modality but practical modality: the field appears as a structured space of possible action relative to a system's self-maintenance constraints.

2 The distinction between orientation and concern marks the boundary between modal differentiation in the thin sense (any system that differentiates a field relative to its structure) and the thick sense (a system whose field-differentiation is governed by its own persistence conditions). Hans Jonas (1966) argued that metabolism is the key to understanding the emergence of concern in living systems: because the organism must constantly replace its own material constituents, it stands in a relation of needful freedom to its environment – dependent on a world it must actively engage. Jonas's insight informs the present account, though MEOW formalises it differently. The 'cost' of self-maintenance – the thermodynamic work the system must perform to preserve its organisation – is what converts mere orientation into genuine concern, and mere constraint-responsiveness into salience.

Different organisms generate different modal differentiations – different salience structures, different affordance gradients, different constraint-profiles – because their self-maintaining organisations are different. A bat and a human occupy the same constraint-field. They generate radically different modal differentiations of it because their mediational architectures differ.

3.3 Self-Maintenance and the Generation of Perspective

What gives rise to modal differentiation? The answer, stated abstractly, is self-maintenance. The answer, stated concretely, is that certain process-clusters within the interactional field are organised such that they actively maintain their own structural coherence against perturbation.

A living organism is the paradigmatic case. An organism is not a static object. It is a dynamically maintained constraint-pattern – a process-cluster that continuously repairs, regulates, and sustains itself through metabolic, homeostatic, and adaptive activity. This self-maintenance is what distinguishes an organism from a boulder. Both are constraint-clusters. Both are scale-dependent stabilisations. But the boulder's stability is passive – it persists through inertia, through the slowness of the geological processes that will eventually erode it. The organism's stability is active – it persists by doing work, by expending energy to maintain its organisation against the thermodynamic gradient that would otherwise dissolve it.

This active self-maintenance generates perspective – not as a metaphysical addition to the organism but as a structural consequence of its organisation. A system that must maintain itself has, by that fact, a differential relationship to its environment. Some interactions support its maintenance; others threaten it. Some constraint-patterns are navigable; others are obstructive. The field, relative to this system, is no longer symmetrical. It has acquired salience, valence, and directionality.

This is the origin of perspective within MEOW. Perspective is not a property of a mental substance looking out at the world. It is the asymmetry generated when a self-maintaining system differentiates the constraint-field relative to its own persistence conditions. The subject does not have a perspective. The subject is the perspective – the organisational profile through which the field becomes directionally structured.[3]

The formulation is substrate-neutral. Biological organisms are the paradigmatic case because biological self-maintenance is the most robust and well-understood form of active constraint-preservation. But nothing in the definition restricts modal differentiation to biological substrates. Any system that exhibits self-maintaining, constraint-responsive organisation – that actively works to preserve its own structural coherence – would generate modal differentiation within the field. Whether current artificial systems exhibit this organisation is an empirical question, not a definitional one. The criterion is organisational, not material:[4] does the system generate the kind of active self-maintenance that differentiates the constraint-field into salience and affordance? If so, it is a locus of modal differentiation. If not, it is not. Three conditions fence off trivial cases. Self-maintenance must be intrinsic – the system bears the cost of maintaining its own organisation, rather than being externally propped up. Constraint-responsiveness must be integrated – woven into the system's organisational coherence, not bolted on as an external controller. And the system must exhibit a unitary salience profile – a

---

3 This formulation inverts the standard relationship between subject and perspective. In the Cartesian and broadly substance-ontological tradition, a subject possesses a perspective – perspective is a property of an independently existing mind. Heidegger's (1927/1962) analysis of Dasein as being-in-the-world offers a structural precursor to the present account: Dasein is not a subject that subsequently acquires a world but a mode of being that is always already worldly. MEOW shares the structural move – the refusal to separate subject from world and then ask how they reconnect – but grounds it in encounter-organisation rather than existential ontology. The identification of subject with perspective (rather than subject as possessor of perspective) is a consequence of taking relational primacy seriously: if encounter is primitive, there is no subject prior to the encountering.

4 The substrate-neutrality of modal differentiation has direct implications for the philosophy of artificial intelligence, though the present chapter does not develop them. If subjecthood is an organisational achievement rather than a material endowment, then the question 'can a machine be a subject?' becomes 'can a machine exhibit self-maintaining, constraint-responsive organisation that generates modal differentiation of the field?' The three conditions stated in the main text (intrinsic self-maintenance, integrated constraint-responsiveness, unitary salience profile) function as structural criteria rather than definitional barriers. Current large language models, for instance, do not bear the cost of their own persistence and do not generate a unitary salience profile from organisational need; they are externally maintained and lack the self-other differentiation that grounds concern. This is an empirical observation about current systems, not a principled exclusion of future ones. Chapter 10 returns to the AI-mind question in the context of the apophatic mind.

coherent centre of valuation from which the field is asymmetrically organised, not a distributed governance regime without a locus of orientation.

3.4 The 'I' as Indexical Compression

If subjecthood is modal differentiation rather than substance, what becomes of the first person? What is the 'I' that appears in every sentence a subject utters about itself?

Within the MEOW framework, the 'I' is an indexical compression – a linguistic device that compresses the entire modal differentiation of a self-maintaining process cluster into a single referential token. When a speaker says 'I see the tree', the word 'I' does not refer to a substance – a soul, a self, or other metaphysical stand-ins – standing behind the encounter. It indexes the modal differentiation from which the encounter is organised: [5]this particular salience structure, this particular mediational architecture, this particular locus of self-maintaining activity within the constraint-field.

The compression is enormous. Everything that constitutes the speaker's organisational profile – biological structure, cognitive history, perceptual bandwidth, experiential trajectory, cultural formation – is packed into two letters. The compression enables linguistic coordination: other agents can identify the speaker as a distinct locus of encounter without needing to reconstruct the full modal differentiation that the word indexes. 'I' is, in the terminology of Chapter 7, a maximally compressed stabilisation – a word that transmits a reference to an entire organisational profile at minimal bandwidth cost.

The compression can also fail, and the failure is instructive. 'I was here yesterday' scrawled on a wall is an orphaned indexical – the compression has outlived the modal differentiation it indexed, and the reference collapses.

5 The treatment of 'I' as indexical rather than referential has antecedents in analytic philosophy of language. Anscombe (1975) argued that 'I' is not a referring expression at all – that it does not pick out an object in the way that names or demonstratives do. Perry (1979) showed that indexical self-reference is essential and irreducible: no amount of third-person description can replace what 'I' does. MEOW's 'indexical compression' draws on both insights but goes further. It agrees with Anscombe that 'I' does not refer to a substance, and with Perry that the indexical function is irreducible. But it adds a diagnostic layer: the reason 'I' appears to refer to a substance is that the grammatical role it occupies (subject position in a subject-predicate sentence) imposes a substance-ontological reading on what is structurally an indexing operation. The illusion of substance-reference is a grammatical artefact, not a philosophical discovery.

'I' in advanced dementia points toward an organisational continuity that is disintegrating; the word persists but the modal differentiation it compresses is fragmenting beneath it. 'I' in dissociative states indexes a differentiation that has temporarily fractured into competing salience profiles, each claiming the pronoun. These are not curiosities. They are evidence that the 'I' is an anchoring practice sustained by organisational coherence, not a label affixed to a metaphysical constant.

But the compression has a cost. It makes the subject look like a thing. The word 'I' is a noun. Grammar places it in subject position: 'I think', 'I perceive', 'I decide'. The subject-predicate structure presupposes a substance bearing properties and performing actions. Every time a speaker uses the word 'I', the ontological grammar of the language reinstalls the very substance architecture that MEOW rejects. The subject is grammatically presented as a thing that exists prior to its encounters and subsequently engages in them. The modal differentiation – the relational, processual, dynamically maintained organisational profile – is compressed into a noun and thereby reified.

This is not a defect of the framework. It is an instance of the ontological grammar thesis developed in Chapter 7. The grammatical medium through which subjecthood is articulated imposes substance-ontological commitments on the articulation. Speakers can think relationally about the subject, but they must express that thinking through a grammar that converts it back into a substance. The 'I' is therefore simultaneously indispensable (as a coordination device) and misleading (as an ontological signal). Using it is a structural necessity. Mistaking it for a description of what the subject fundamentally is – a thing, a substance, a metaphysical atom – is an error generated by the encoding protocol.

3.5 The 'Self' as Diachronic Narrative Compression

If the 'I' compresses modal differentiation at a moment, the 'self' compresses it across time.

The self, within this framework, is a diachronic narrative compression – a stabilised pattern that tracks organisational continuity across temporally extended sequences of encounter-events. When a person says 'I am the same person I was ten years ago', the claim is not that a substance has persisted unchanged through a decade of experience. It is that the modal differentiation has exhibited sufficient organisational continuity – continuity of memory, of embodied structure, of dispositional profile, of social role – to sustain a compressed narrative of persistence.[6]

The self is therefore a heuristic. It is not fictitious – the organisational continuity it tracks is real. Memories are real. Dispositional patterns are real. The embodied structure that carries them persists through metabolic renewal with sufficient stability to ground re-identification. But the 'self' as a unified, bounded, persistent entity – the thing that the word appears to name – is an abstraction imposed on this continuity by the same grammatical and cognitive machinery that converts processes into objects throughout the rest of the framework.

The heuristic is extraordinarily useful. Social life depends on it. Moral responsibility, legal liability, personal relationships, institutional roles, and political participation all presuppose that persons persist through time as identifiable agents. The self-heuristic enables this presupposition to function. Without it, the coordination mechanisms of Chapters 7 and 8 would have nothing stable to coordinate. The 'self' persists both narratively (first-person continuity through memory and identity-feel) and institutionally (third-person re-identification through records, roles, and liability), and the two need not coincide. A person with amnesia retains institutional continuity while narrative continuity fractures. A person who has radically changed their values retains both, despite the felt discontinuity. The divergence between narrative and institutional compression is precisely where the social ontology of Chapter 8 bites: institutions track a self that may not match the self the subject experiences.

6 The treatment of personal identity as organisational continuity rather than substance-persistence places MEOW in the neighbourhood of neo-Lockean psychological continuity theories (Locke, 1690; Parfit, 1984), which ground identity in overlapping chains of memory and psychological connectedness rather than in the persistence of a soul or body. The structural parallel is real: both traditions reject substance as the criterion of identity. But the divergence is significant. Psychological continuity theories remain within a broadly subject-centred framework – the question is what makes this person the same person over time. MEOW reframes the question: there is no 'person' as a primitive entity whose identity is at issue. There is a modally differentiated process-cluster whose organisational continuity is tracked by the heuristic we call 'self'. The identity question is not answered but dissolved – replaced by a question about the stability conditions of a dynamic pattern. Parfit's conclusion that personal identity is 'not what matters' is structurally consonant with this dissolution, though his reasons differ.

But the heuristic is also a source of philosophical confusion when it is mistaken for a metaphysical discovery. The persistence of the self feels like the persistence of a substance – an enduring thing that bears changing properties. The feeling is partly experiential (memory creates the sense of continuity), partly grammatical (the word 'I' functions as a noun across temporal contexts), and partly institutional (social systems reinforce the fiction of personal identity through legal and administrative mechanisms). These reinforcing pressures make the substance reading of the self feel inevitable. Understanding that it is a compression rather than a substance does not eliminate the feeling. It does explain why the feeling persists even after the metaphysics has been revised.

3.6 Agency Without Substance

Substance ontology grounds agency in the causal powers of an independently existing agent. The agent decides, the agent acts, the agent is responsible. The metaphysical picture is of a self-contained entity that initiates causal sequences through the exercise of will.

MEOW reframes agency without this metaphysical picture.

Agency, within this framework, is the capacity of a modally differentiated system to alter its own encounter-trajectory in response to its salience structure. It is not the exercise of a hidden causal power by a substance. It is a structural feature of sufficiently complex self-maintaining systems: they do not merely undergo encounters passively but actively shape the encounters they participate in, selecting among available affordance gradients, modifying their mediational organisation, and altering the constraint-patterns of subsequent encounters.

The reframing preserves everything that matters about agency while discarding the metaphysical surplus. A person who decides to cross the street is a modally differentiated system that has assessed the affordance gradient (the road is crossable, the traffic permits passage), integrated that assessment with its persistence conditions (it needs to reach the other side), and initiated a sequence of encounter-events that alters its trajectory through the constraint-field. Nothing in this description requires a Cartesian ego, an uncaused cause, or a metaphysical substance exercising will. What it requires is a self-maintaining system with sufficient organisational complexity to assess, integrate, and act – which is what biological organisms do.

The question of free will – whether the agent's decision is 'truly free' in some metaphysical sense – does not arise within this framework in its traditional form. The traditional form presupposes substance ontology: a substance must be the ultimate originator of its actions for those actions to be 'free'. MEOW does not posit such substances. What it posits is modally differentiated systems operating within constraint-fields, whose actions are neither fully determined by prior encounter-sequences (because the system's self-maintaining organisation introduces organisation-dependent flexibility – sensitivity to higher-order structure such as models, forecasts, and learned priors that re-shapes the trajectory space) nor uncaused (because every action occurs within a constraint-context that shapes its possibility space). The framework is compatible with compatibilist accounts of freedom but does not require them. The question of freedom becomes a question about the organisational profile of the agent – how much structural flexibility the system's self-maintenance affords – rather than a question about the metaphysical status of causation.[7]

3.7 Recursion: Subjects Who Theorise About Subjecthood

A framework that accounts for subjecthood must be able to account for itself. If the subject is a modally differentiated process-cluster within the constraint-field, then the philosopher writing about subjecthood is also a modally differentiated process-cluster – one that happens to be directing its organisational complexity toward the articulation of its own structure.

Does this create a vicious regress? Can the framework account for its own production?

The answer is that it can, and that doing so is not as exotic as it sounds.

7 The reframing of freedom as organisational flexibility rather than metaphysical origination sidesteps the traditional free will debate (libertarianism vs. determinism vs. compatibilism) without engaging it on its own terms. This is deliberate. The traditional debate is formulated within substance ontology: it asks whether a substance (the agent) can be the uncaused initiator of a causal chain. MEOW does not posit such substances, so the question as traditionally framed does not arise. What replaces it is a graduated account: systems with greater organisational complexity – richer models of their own constraint-field, wider counterfactual sensitivity, deeper recursive integration – exhibit greater freedom in the structural sense that they can navigate a wider affordance space. This is closer to Dennett's (1984, 2003) 'degrees of freedom' approach than to classical compatibilism, though MEOW does not require Dennett's functionalist commitments.

A modally differentiated system is, by definition, a system capable of reflexive integration – of incorporating its own prior states into its current encounter-organisation. This reflexive capacity is what distinguishes phenomenal presentation from mere presentational structure (Chapter 1, Section 1.6). A system that can integrate its own prior states can also, given sufficient complexity, take its own organisation as an object of encounter. It can model its own mediational architecture. It can notice – imperfectly, through the same finite-bandwidth protocol that constrains all articulation – the structure of its own modal differentiation.

Philosophy is what happens when this capacity is directed toward the most general features of encounter-organisation. Or, stated less charitably: philosophy is the attempt to say 'process' in a language that keeps insisting on 'things'. The philosopher who writes about subjecthood is a self-maintaining process-cluster whose reflexive integration has become sufficiently articulated to take the structure of subjecthood itself as an object of encounter. The articulation is conducted through the synchronisation protocol (language), which imposes its own ontological grammar on the result. The philosopher therefore confronts the predicament described in Chapter 7: the medium of articulation imposes substance-ontological commitments on the content being articulated. Every sentence about modal differentiation is grammatically converted into a sentence about a thing.

This is not a refutation of the framework. It is a prediction of the framework. The ontological grammar thesis predicts that any articulation of non-substance ontology conducted in a substance-grammar language will face articulatory resistance. The philosopher's experience of that resistance – the feeling that relational accounts of the self are 'awkward' or 'counterintuitive' – is itself evidence for the thesis. The awkwardness is not a sign that the framework is wrong. It is a sign that the encoding protocol is taxing the articulation, exactly as Chapter 7 predicts.

3.8 The Subject Within Language and Institutions

The account of subjecthood developed in this chapter does not stand alone. It connects directly to the accounts of language (Chapter 7) and institutions (Chapter 8), forming a triangle – Subject–Language–Institution – that constitutes the coordination stack the remainder of the monograph presupposes.

Subjects are what language synchronises. The stabilisations that the synchronisation protocol compresses and transmits are stabilisations of modally differentiated systems. When one agent communicates with another, what is being aligned is not raw data but oriented encounter-patterns – patterns organised by each agent's modal differentiation, shaped by each agent's salience structure, compressed through each agent's mediational architecture. The imperfections of synchronisation identified in Chapter 7 are, at root, imperfections in the alignment of differently differentiated encounter-organisations. Two agents with identical modal differentiations (were such a thing possible) would achieve perfect synchronisation. Two agents with radically divergent differentiations would achieve none. Actual communication falls somewhere between these limits, along the gradient described in Figure 7.1. Within relatively homogeneous in-groups – agents sharing mediational architecture, cultural formation, and ontological grammar – decompression variance shrinks and synchronisation approaches the upper bound. Across pluralistic populations, ontological variance dominates and pulls the curve down.

Subjects are what institutions constrain. The second-order constraint systems described in Chapter 8 operate on modally differentiated agents. Laws, norms, and institutional roles alter the affordance gradients available to subjects – they change what the self-maintaining system can do, must do, and is prevented from doing. Institutional constraint is not imposed on inert objects. It is imposed on systems that actively assess, respond to, and sometimes resist the constraints placed upon them. The dynamic between institutional constraint and agentive response is what gives social ontology its characteristic complexity: institutions constrain agents who are themselves capable of modifying, reinterpreting, or undermining the constraints.

Subjects are where ontological grammar becomes invisible. The ontological commitments of the synchronisation protocol are most consequential – and least visible – within the subject's own self-understanding. When a person says 'I have a mind' or 'I am a self', the grammatical structure of the sentence installs a substance that the framework rejects. The subject's self-model is conducted in the same noun-dominant grammar that generates the substance ontology of the broader philosophical tradition. This means that the most intimate aspects of self-understanding – the sense of being a thing, of having an inner life, of persisting as an entity through time – are partly artefacts of the encoding protocol rather than transparent reports of metaphysical reality.

This is not eliminativism. The experience of selfhood is real. The organisational continuity that the self-heuristic tracks is real. The modal differentiation that generates perspective is real. What is not real – or rather, what is heuristic rather than fundamental – is the substance that grammar projects behind these organisational features. The self is a genuine pattern. The soul is a grammatical shadow.

3.9 The Eliminativist Worry

A concern will have been forming in the reader's mind, and it deserves direct address.

If the subject is not a substance but a modal differentiation, and if the self is a heuristic compression rather than a metaphysical entity, then has the framework eliminated the subject? Has it explained subjecthood away?

No. And the distinction matters.

Eliminativism denies that the phenomenon exists. It says: there are no subjects, no selves, no agents – only physical processes misdescribed by folk psychology. MEOW says something different. It says: subjects, selves, and agents are real – they are genuine organisational features of the constraint-field – but they are not the kind of thing that substance ontology says they are. They are not substances. They are not metaphysical atoms. They are not independently existing things that subsequently enter into relations. They are relational patterns that emerge from the organisation of encounter and are sustained by active self-maintenance.

The distinction between 'it doesn't exist' and 'it exists but is not what you thought it was' is the distinction between eliminativism and reconceptualisation. MEOW reconceptualises subjecthood. It does not eliminate it.[8]

8 The eliminativist–reconceptualist distinction is more contested than the main text's confident formulation might suggest. Dennett (1991) has been read both as an eliminativist about consciousness (there is no 'Cartesian theatre' and therefore no consciousness as traditionally conceived) and as a reconceptualist (consciousness exists but is a 'fame in the brain' rather than a unified inner show). The Churchlands' eliminative materialism denies the ontological adequacy of folk-psychological categories altogether. MEOW's position is that the organisational features that folk psychology tracks – perspective, persistence, agency – are real, even though the ontological interpretation that folk psychology and substance ontology impose on them (a thing behind the features) is not. This places the framework

The test is practical. Does the framework preserve the features of subjecthood that matter – perspective, persistence, agency, moral responsibility, the capacity for self-reflection? It does. Each of these features has been given a structural account within the architecture of modal differentiation. None has been denied. What has been denied is only the metaphysical interpretation that substance ontology places on these features – the claim that they require a self-subsisting thing behind them. They do not. They require an organisational profile. That profile is real, and it is what we are.

3.10 What This Chapter Has Established

Subjecthood is modal differentiation – the point at which the interactional field becomes asymmetrically organised relative to a self-maintaining, constraint-responsive process cluster. The subject is not outside the field. It is a dynamic differentiation within it, generated by the organisational complexity of the self-maintaining system.

The formulation is substrate-neutral: biological organisms are the paradigmatic but not the sole possible locus of such differentiation. The 'I' is indexical compression – a linguistic device that packs the full modal differentiation into a single referential token, at the cost of grammatically reinstalling the substance ontology the framework rejects. The 'self' is diachronic narrative compression – a stabilised pattern tracking organisational continuity across time, useful as a heuristic but misleading when mistaken for a metaphysical substance. Agency is the capacity of a modally differentiated system to alter its own encounter-trajectory in response to its salience structure.

The framework is not eliminativist. It preserves the features of subjecthood that matter – perspective, persistence, agency, self-reflection – while denying that they require a substance behind them. The subject is a genuine pattern within the constraint-field. The soul is a grammatical shadow cast by the encoding protocol.

Subjects are what language synchronises, what institutions constrain, and where ontological grammar becomes most consequential and least visible. The triangle of subject, language, and institution – developed across

closer to Dennett's reconceptualist reading than to the Churchlands' eliminativism, but without Dennett's functionalist metaphysics. The test proposed in the main text – whether the framework preserves the features that matter – is deliberately practical rather than metaphysical.

Chapters 3, 7, and 8 – forms the relational architecture within which the remainder of the monograph operates. The chapters that follow develop the remaining structural components: logic, epistemology, perception, the realist defence, and the framework's application and limits.

4

Logic

Coherence Grammar Under Constraint

Contradiction is not a sign of falsity.
La contradiction n'est pas une preuve de fausseté.

— Blaise Pascale, *Pensées*

Tell all the truth but tell it slant.

— Emily Dickinson, *poem #1263*

4.1 Introduction: The Status of Logic

THE PRECEDING CHAPTERS HAVE established an ontology of structured interaction, an account of subjecthood as modal differentiation, and – in the chapter that follows this one in the argumentative sequence – an epistemology grounded in stabilisation under encounter pressure. The question that now intervenes is structural: what conditions must stabilisations satisfy in order to remain internally coherent?

This is the question of logic. And the answer MEOW gives differs from both of the dominant traditions.

The Platonist holds that logical laws are eternal, mind-independent truths – features of an abstract realm that obtains regardless of whether any encountering system exists. On this view, the law of non-contradiction is not a human invention. It is a fact about reality that we discover, as we discover mathematical truths: timelessly, universally, and without dependence on mediation.

The conventionalist holds that logical laws are agreed-upon rules – social or linguistic conventions adopted for their usefulness and revisable in principle. On this view, logic is a tool. It has no deeper grounding than the consensus of those who employ it. Different communities could, in principle, adopt different logics, and none would be more 'correct' than any other.

MEOW rejects both positions and occupies a space that neither anticipates. Logic is neither discovered in a transcendent realm nor invented by consensus. It is a coherence grammar – the minimal structural condition required to preserve consistency in stabilised orientation under constraint.[1] Logic emerges because stabilisations that violate it cannot survive encounter

1 The landscape of positions on the status of logic is considerably more varied than the Platonist–conventionalist dichotomy presented here. Quine (1951) argued that logical truths are revisable in principle, continuous with empirical belief, and that no sharp analytic–synthetic boundary underwrites their alleged necessity. Putnam (1968) went further, asking whether logic might be empirical in the same way that geometry turned out to be. Dewey (1938) treated logic as a theory of inquiry – not a body of eternal truths but a formalisation of the conditions under which inquiry succeeds. Brandom (1994, 2000) grounds logical vocabulary in inferential practice rather than in representation of abstract objects. MEOW's 'coherence grammar' occupies ground adjacent to Dewey and Brandom: it treats logical principles as conditions of successful stabilisation rather than discoveries about a transcendent realm or conventions of a community. The dichotomy in the main text is a pedagogical simplification; the philosophical terrain is more crowded than the chapter's compressed treatment suggests.

pressure. It is not imposed on reality from above (Platonism) or from below (convention). It is extracted from the operating conditions of any system that must maintain stable orientation within a constraint-field.

4.2 Coherence Grammar: The Core Concept

The phrase 'coherence grammar' requires unpacking, because it is carrying the chapter's central claim.

A grammar, in the linguistic sense, is a set of structural rules that determine which combinations of elements are well-formed and which are not. 'The cat sat on the mat' is grammatically well-formed in English. 'Mat the on sat cat the' is not. The grammar does not determine what is true – both 'the cat sat on the mat' and 'the elephant danced on the ceiling' are grammatically well-formed – but it determines what is structurally admissible. It sets the conditions for intelligibility.

Logic, within MEOW, functions analogously but at a different level. Where linguistic grammar determines the structural conditions for well-formed sentences, logical grammar determines the structural conditions for well-formed stabilisations. A stabilisation that affirms and denies the same constraint-pattern simultaneously is not false in the way that 'the cat is on the mat' might be false if the cat is elsewhere. It i2673s incoherent – structurally incapable of maintaining itself under encounter pressure, because the encounter-field cannot simultaneously confirm and disconfirm the same constraint-pattern relative to the same mediational architecture.

The core logical principles – identity, non-contradiction, excluded middle – are therefore not discoveries about an abstract realm. They are stability conditions. They describe the minimum structural requirements that any stabilisation must satisfy if it is to persist as a coherent orientation within the constraint-field.

Identity (a thing is what it is) expresses the requirement that a stabilisation must track the same constraint-pattern across encounters. If 'this boulder' refers to a different constraint-cluster every time it is invoked, the stabilisation dissolves. Identity persistence is the condition for re-identification – for treating a stabilisation as the same stabilisation across temporal and contextual variation.

Non-contradiction (a thing cannot both be and not be the case simultaneously) expresses the requirement that a stabilisation must not simultaneously affirm and negate the same constraint-pattern relative to the same encounter-conditions. A system that stabilises the floor as both weight-bearing and not weight-bearing cannot navigate. The stabilisation offers no orientation. It is not wrong – it is inoperable.

Excluded middle (a thing either is or is not the case) is the most contested of the three, and MEOW treats it with appropriate caution. In domains where constraint-patterns are sharp and well-discretised – the Invariants region – excluded middle holds straightforwardly. The decay constant either has this value or it does not. In domains where constraint-patterns are graded, fluid, or resistant to discretisation – the Contestables and Fluids regions – excluded middle becomes a grammatical imposition rather than a structural necessity. Forcing a binary on a gradient is a compression artefact, not a logical discovery. The principle holds where the domain supports it and strains where the domain resists discretisation.[2] This is not a failure of logic. It is evidence that logical grammar, like linguistic grammar, has operating conditions – and that those conditions are encounter-relative.

4.3 Logic Is Not Platonic

The Platonist account of logic has deep appeal. It explains the apparent necessity of logical truths – they seem to be true in all possible worlds, regardless of what anyone thinks about them – and it explains their apparent universality – they seem to apply to everything, without exception.

MEOW preserves both features without the Platonic metaphysics.

The necessity of logical principles, within the framework, derives not from their existence in an abstract realm but from their status as stability

2 The claim that excluded middle is domain-sensitive rather than universally prescriptive aligns MEOW with the broad family of positions that restrict bivalence. Intuitionistic logic (Brouwer, Heyting) rejects excluded middle on constructive grounds: a proposition is neither true nor false until a proof or refutation is produced. Supervaluationism preserves excluded middle as a logical schema while denying bivalence for borderline cases. Many-valued logics assign intermediate truth-values. MEOW's position is distinct from all of these but closest to the supervaluationist intuition: excluded middle is structurally sound in domains where constraint-patterns support sharp discretisation and structurally misleading in domains where they do not. The framework does not reject excluded middle; it identifies its operating conditions. This is a contextualisation rather than a restriction – a claim about the grammar's domain of application rather than about its validity.

conditions. A stabilisation that violates non-contradiction cannot survive encounter pressure – not because a Platonic law forbids it, but because the constraint-field cannot simultaneously confirm and disconfirm the same pattern. The necessity is structural, not transcendent. It holds in every encounter because it follows from the operating conditions of encounter itself.

The universality of logical principles derives from the generality of those operating conditions. Any system that must maintain coherent stabilisations within a constraint-field – regardless of its biological substrate, its cultural formation, or its mediational architecture – must satisfy the same structural requirements. Non-contradiction is universal not because it governs an abstract realm but because any system that violates it loses the capacity to navigate. The universality is functional, not metaphysical.

The Platonist might object: if logical principles are merely stability conditions, they could in principle be violated by a system operating under different conditions. But this objection misidentifies the claim. MEOW does not say that logical principles are merely stability conditions, as though 'mere' diminishes them. It says that being a stability condition is a sufficient explanation of their necessity and universality. The Platonic realm is explanatorily redundant. Logic's necessity is fully accounted for by the operating conditions of stabilisation under constraint. Adding a transcendent realm of abstract objects adds nothing that the structural account does not already provide.

4.4 Logic Is Not Convention

The conventionalist account also has appeal. It explains the apparent plurality of logical systems – classical, intuitionistic, paraconsistent, many-valued – and it explains why logical practice seems to vary across contexts.

MEOW again preserves the legitimate insight whilst rejecting the metaphysical conclusion.

The insight is correct: logical systems are plural. Different formalisms encode different structural requirements, and the choice among them is partly determined by the domain of application and the purposes of the enquiry. Intuitionistic logic is appropriate for constructive mathematics. Paraconsistent logic is appropriate for contexts where contradictions must be tolerated without explosion. Fuzzy logic is appropriate for domains where vagueness is structural rather than eliminable.

But the plurality of logical systems does not entail the conventionality of logical principles. The various formalisms are different encodings of coherence requirements, tailored to different encounter-conditions. They are not arbitrary choices among equally valid options. They are specialised grammars for different domains. Classical logic encodes the coherence requirements appropriate to well-discretised, binary-valued domains. Intuitionistic logic encodes those appropriate to constructive domains where existence requires proof. Each is constrained by the encounter-conditions of the domain it serves.

The conventionalist's error is to confuse the plurality of encodings with the arbitrariness of the requirements.[3] Different grammars are possible. The underlying structural demand – that stabilisations be coherent enough to survive encounter pressure – is not conventional. It is a consequence of the operating conditions of any system that must maintain orientation within a constraint-field.

4.5 Logic and Ontological Grammar

Chapter 7 develops the claim that the grammatical structure of natural language imposes ontological commitments on everything it encodes. Logic – particularly classical logic – exhibits a parallel entanglement.

Classical logic is built on subject-predicate structure. A proposition asserts that a subject has a property or stands in a relation: 'Socrates is mortal', 'the electron is charged', 'A is greater than B'. The logical form presupposes entities (subjects) that bear properties (predicates). This is the same substance-ontological architecture that Indo-European grammar installs at the level of syntax.

3 The distinction between plurality of encodings and arbitrariness of requirements corresponds, at a structural level, to the distinction between logical pluralism and logical relativism. Beall and Restall (2006) defend a pluralism in which multiple logics can be simultaneously correct because 'validity' is legitimately precisified in different ways. This is compatible with MEOW's treatment: different coherence grammars are appropriate to different encounter-conditions, and the plurality reflects genuine structural variation in the domains they serve. What the framework resists is the further step to relativism – the claim that the choice among logics is unconstrained by the encounter-conditions of the domain. Plurality is structural; arbitrariness is not. The coherence grammar appropriate to a domain is determined by the domain's constraint-structure, not by the preferences of the logician.

The parallel is not accidental. Classical logic was formalised within and for a philosophical tradition that assumed substance ontology. Its propositional structure reflects the same ontological commitments that the grammar of the language in which it was developed presupposes.[4] Subject-predicate logic is the coherence grammar appropriate to a substance-ontological framework. It encodes the structural requirements for stabilisations organised around objects bearing properties.

MEOW does not reject classical logic. It contextualises it. Subject-predicate logic is the appropriate coherence grammar for domains where substance-shaped stabilisations are well-aligned with the underlying constraint-patterns – where the phenomena genuinely behave like discrete objects bearing determinate properties. In the Invariants region, this alignment is typically strong. Classical logic works well for physical constants, well-individuated objects, and binary-valued propositions precisely because the domain's constraint-structure matches the logic's ontological grammar.

In domains where the constraint-patterns are processual, relational, or graded – where the phenomena resist discretisation into subjects and predicates – classical logic's ontological grammar becomes a source of distortion. Forcing a relational process into subject-predicate form introduces the same kind of categorial misalignment that Chapter 7 identifies in natural language. The logical form imposes a structure that the domain does not support, and the resulting formalisation is grammatically well-formed but ontologically misleading.

This is where alternative logics earn their place – not as conventions adopted for convenience but as coherence grammars appropriate to different encounter-conditions. Process logic, relational logic, and graded-truth logics

4 The claim that classical logic's subject-predicate structure reflects and reinforces substance ontology is not new. Peirce's development of relational logic was partly motivated by the recognition that syllogistic logic's subject-predicate form could not adequately represent relational propositions. Frege's function-argument analysis replaced the subject-predicate structure with a more general framework, though Frege himself did not draw the ontological conclusion that MEOW draws. More recently, process logicians have argued that standard predicate logic is ill-suited to processual ontologies, precisely because its quantificational structure presupposes a domain of persistent objects. MEOW's contribution is to treat this observation not as a curiosity in the history of logic but as a diagnostic: the alignment between logical form and ontological commitment is itself a case of the ontological grammar thesis developed in Chapter 7. Logic does not merely represent a substance ontology; its standard formalisation enacts one.

are attempts to encode coherence requirements for domains that substance-shaped classical logic distorts. Their proliferation is not evidence of logical conventionalism. It is evidence that different domains require different coherence grammars, just as different organisms generate different modal differentiations of the same constraint-field.

4.6 Quantum Mechanics and the Limits of Object-Grammar

An illustrative case – treated briefly, not as the chapter's centrepiece – shows the ontological grammar of logic reaching its limits.

Quantum mechanics notoriously resists classical logical structure. Superposition, entanglement, and the measurement problem all strain the subject-predicate architecture of classical logic. An electron in superposition does not have a determinate position until measured. Asserting 'the electron is at position x' or 'the electron is not at position x' presupposes that the electron is the kind of entity to which determinate position predicates apply. Superposition suggests it is not – or at least, not prior to the encounter-event constituted by measurement.

The standard response is to say that quantum mechanics 'breaks' classical logic – that quantum phenomena require a new logic (quantum logic) in which distributivity fails and the classical rules no longer hold.[5]

MEOW reframes the situation. Quantum mechanics does not break logic. It reveals the scale limits of object-grammar. Classical logic – with its subject-predicate structure, its bivalence, its assumption of determinate property-attribution – is the coherence grammar appropriate to a domain in which constraint-patterns are object-shaped: discrete, bounded, and

5 The proposal that quantum mechanics requires a revision of logic originates with Birkhoff and von Neumann (1936), who showed that the lattice of subspaces of a Hilbert space does not satisfy classical distributivity. Putnam (1968) drew the stronger conclusion that logic might be empirical – revisable in light of physical evidence. The subsequent debate has been extensive and inconclusive: some philosophers of physics accept quantum logic as a genuine logical reform; others treat it as a reformulation of quantum theory's mathematical structure with no implications for logic as such. MEOW's position is orthogonal to this debate. The framework does not ask whether quantum mechanics forces a change of logic. It asks why classical logic strains in the quantum domain and answers: because the domain's constraint-structure is not object-shaped, and classical logic's ontological grammar presupposes that it is. The problem is grammatical mismatch, not logical breakdown.

property-bearing. At the quantum scale, the constraint-patterns are not object-shaped. They are relational, probabilistic, and context-dependent in ways that the object-grammar of classical logic cannot accommodate without distortion.

The 'paradoxes' of quantum mechanics are therefore not logical paradoxes. They are grammatical mismatches – cases in which a coherence grammar designed for one scale of encounter is applied to a domain whose constraint-structure does not support it. The electron is not defying logic. The logic is being asked to describe a domain for which its ontological grammar was not built.

This illustrative case is not meant to resolve the interpretation of quantum mechanics – that lies far outside this monograph's scope. It is meant to demonstrate a general principle: logical grammar, like linguistic grammar, has operating conditions. When those conditions are met, the grammar provides coherent structure. When they are not, the grammar produces artefacts that are mistaken for deep mysteries. The 'mystery' of quantum logic is an artefact of applying object-grammar to a non-object domain. The mystery dissolves when the grammar is recognised as encounter-relative rather than universally prescriptive.

4.7 Inference as Constraint-Transmission Within Stabilisations

If logic provides the coherence grammar for stabilisations, inference is the mechanism through which coherence is maintained across sequences of stabilisations.

Inference, within this framework, is constraint-transmission within networks of stabilisations. When a system moves from one stabilisation to another – from 'this substance is toxic' to 'I should not ingest it' – the transition is constrained by the coherence grammar. The conclusion must be structurally compatible with the premise. The constraint-pattern encoded in the premise must be preserved in the conclusion. If the transition violates the coherence requirements – if it affirms what the premise denies, or denies what the premise entails – the resulting stabilisation is incoherent and cannot survive encounter pressure.

Deductive inference is the limiting case: constraint-transmission with no loss. If the premises are stable and the coherence grammar is satisfied, the conclusion inherits the stability of the premises. This is why deductive reasoning feels necessary – because the constraint is fully transmitted, and any violation of the conclusion would simultaneously violate the premises.

Inductive inference involves constraint-transmission with compression. The move from 'every observed instance of X has been followed by Y' to 'the next instance of X will be followed by Y' compresses a finite set of encounter-events into a generalised stabilisation. The compression introduces risk – the next encounter may not cooperate – but the stabilisation is constrained by the pattern of prior encounters. Induction is not logically guaranteed. It is encounter-pressured. Its reliability depends on the stability of the underlying constraint-patterns.

Abductive inference – inference to the best explanation – involves selecting among competing stabilisations on the basis of their coherence, scope, and constraint-responsiveness.[6] It is the epistemic equivalent of natural selection among stabilisations: the stabilisation that best accounts for the widest range of encounter-patterns, with the least internal incoherence and the greatest constraint-sensitivity, is provisionally adopted. This is not a deductive guarantee. It is a pragmatic optimisation under the coherence grammar.

---

6 The treatment of inference as constraint-transmission has a structural affinity with Brandom's (1994) inferentialism, which grounds the content of concepts in their inferential roles rather than in their representational correspondence to objects. For Brandom, to grasp a concept is to be able to navigate the web of inferences in which it participates. MEOW's account is adjacent: to stabilise a constraint-pattern is to be positioned within a network of constraint-transmissions whose coherence grammar determines what follows from what. The divergence is that Brandom's inferentialism operates within a broadly discursive framework – inference is a practice of giving and asking for reasons among language-users – whereas MEOW's constraint-transmission is pre-discursive. A thermostat transmits constraint. A plant's phototropism transmits constraint. Discursive inference is the linguistically mediated special case, not the paradigm.

Purpose: Engaging inferentialism. Brandom is named in the Author Position Statement as a structural influence, but the chapter on logic never cites him. Since his inferentialism is the most prominent contemporary account of the relationship between inference and content, the omission is conspicuous. The footnote marks the structural kinship and the divergence (pre-discursive constraint-transmission vs. discursive scorekeeping). Strong candidate.

Each form of inference is a mode of constraint-transmission, differentiated by how much compression is involved and how much encounter-pressure the result can withstand. Logic provides the grammar that all three must satisfy. Epistemology (Chapter 5) provides the account of what makes the resulting stabilisations knowledge rather than speculation.

4.8 What This Chapter Has Established

Logic is a coherence grammar – the minimal structural condition required to preserve consistency in stabilised orientation under constraint. It is neither Platonic (existing independently of all encounter in an abstract realm) nor conventional (adopted by agreement and revisable at will). It is a stability condition extracted from the operating requirements of any system that must maintain coherent orientation within a constraint-field.

The core logical principles – identity, non-contradiction, and (with qualification) excluded middle – are structural requirements for the persistence of stabilisations under encounter pressure. Their necessity is structural, not transcendent. Their universality is functional, not metaphysical. The Platonic realm is explanatorily redundant. Convention is explanatorily insufficient.

Classical logic's subject-predicate structure reflects and reinforces the substance-ontological commitments of the language and tradition in which it was formalised. It is the appropriate coherence grammar for well-discretised, object-shaped domains. In domains where constraint-patterns are processual, relational, or graded, classical logic's ontological grammar becomes a source of distortion. Alternative logics are not conventions but specialised coherence grammars for different encounter-conditions. The proliferation of logical systems is evidence of domain-sensitivity, not arbitrariness.

Inference is constraint-transmission within networks of stabilisations, differentiated by compression level and encounter-pressure tolerance. Logic provides the grammar; epistemology provides the justification; encounter provides the test.

The architecture now has its formal structure: encounter-events (Chapter 1), constraint and ontology (Chapter 2), subjecthood (Chapter 3), logical coherence (this chapter), and epistemological justification (Chapter 5). What remains to be developed is the account of how the constraint-field becomes structured experience – how organisms navigate, perceive, and orient themselves within the field through affordance gradients. That is the subject of Chapter 6.

5

Epistemology

Convergence, Error, and the Structure of Justification

The opinion which is fated to be ultimately agreed to by all who investigate is what we mean by the truth.

— Charles S Peirce

A map is not the territory it represents, but, if correct, it has a similar structure to the territory, which accounts for its usefulness.

— Alfred Korzybski

5.1 The Epistemic Problem

THE PRECEDING CHAPTERS HAVE established what exists (structured interaction), what agents are (modally differentiated process-clusters), and how the constraint-field is organised relative to self-maintaining systems. The question that now presses is: what does it mean, within this architecture, to know something?

The question is urgent because the framework has been systematically removing the traditional supports of epistemology. There is no unmediated access to a mind-independent world. There is no representational mirror in which reality is faithfully reflected. There are no epistemic foundations – no bedrock of certainty upon which the rest of knowledge can be built. Mediation is constitutive. Compression is unavoidable. The synchronisation protocol is finite-bandwidth and ontologically non-neutral.

A reader could be forgiven for wondering whether the framework has made knowledge impossible.

It has not. It has made a particular picture of knowledge impossible – the picture in which knowing consists of possessing an accurate internal representation of an external world. That picture depended on the substance ontology MEOW rejects: a knowing subject on one side, a known world on the other, and knowledge as the bridge between them. Once the bridge architecture collapses, a different account of knowledge is needed.

MEOW provides one. Knowledge is not representation. It is stabilisation under encounter pressure.[1] A system knows something when it has formed a stabilisation of the constraint-field that survives the ongoing pres-

1 The replacement of representation with stabilisation places MEOW within a broad anti-representationalist current in epistemology and philosophy of mind, though the framework's specific formulation differs from each of its neighbours. Rorty (1979) argued that the entire metaphor of mind as mirror should be abandoned; his alternative was a conversational pragmatism without a positive epistemological replacement. Hacking (1983) distinguished representing from intervening, arguing that experimental engagement with the world provides epistemic traction that representational correspondence cannot. The enactivist tradition (Varela, Thompson, & Rosch, 1991) grounds knowledge in embodied sense-making rather than internal modelling. MEOW shares the anti-representational commitment of all three but offers a structural account that none of them provides: stabilisation under encounter pressure is a mechanism, not merely a programme. It specifies how knowledge emerges (through constraint-testing), why it tends toward reliability (through the veto structure of encounter), and where it fails (through compression loss, modal divergence, and ontological grammar distortion).

sure of encounter – when its orientation within the field is reliable enough to support navigation, prediction, and coordination. Knowledge is not a picture of reality. It is effective engagement with reality's constraints.

5.2 Encounter Pressure and Epistemic Selection

The mechanism through which knowledge emerges is encounter pressure.

Encounter continuously subjects stabilisations to testing. Every prediction that fails, every affordance that misfires, every expectation that collapses is an instance of the constraint-field refusing to cooperate with the stabilisation that has been imposed on it. A stabilisation that predicts the floor will support your weight is tested every time you step. A stabilisation that categorises a substance as edible is tested every time you eat. A stabilisation that expects a social norm to hold is tested every time you act on the expectation.

Stabilisations that survive this pressure persist. Stabilisations that do not are revised or abandoned. Knowledge therefore behaves like a selection process: stabilisations are formed through encounter, tested against ongoing constraint-pressure, and filtered by the field's refusal to accommodate those that are misaligned.[2]

This is why knowledge tends toward reliability even though mediation never disappears. The world does not transmit truth to the knower. It does something more modest and more powerful: it vetoes bad compressions. Reality's primary epistemic contribution is not revelation but refusal – the persistent inability of poorly aligned stabilisations to survive encounter.

2 epistemology treats scientific conjectures as analogues of biological organisms: they are generated, tested against reality, and eliminated when they fail. Campbell (1974) developed a 'blind variation and selective retention' model of knowledge-growth. Toulmin (1972) applied evolutionary reasoning to the development of concepts and intellectual traditions. MEOW's 'encounter pressure' is structurally adjacent to Popper's falsification pressure but differs in two respects. First, the selection mechanism is not limited to propositional conjectures; it operates on all stabilisations, including perceptual, practical, and institutional ones. Second, the mechanism is not merely eliminative (vetoing bad stabilisations) but also formative: encounter pressure shapes stabilisations as they are formed, not only after they have been articulated. The constraint-field is not a passive tribunal that accepts or rejects finished theories. It is an active participant in the stabilisation process.

The mechanism is neither foundationalist nor coherentist in the traditional sense, though it shares features with both.[3] It is not foundationalist because there are no unmediated epistemic foundations – no privileged class of beliefs that are directly justified by contact with reality. Every stabilisation is mediated. It is not purely coherentist because coherence alone is insufficient – a set of stabilisations can be internally coherent and yet comprehensively misaligned with the constraint-field. What MEOW adds is constraint-responsiveness: the demand that stabilisations not merely cohere with one another but survive ongoing encounter pressure. Coherence without constraint-responsiveness is fantasy. Constraint-responsiveness without coherence is chaos. Knowledge requires both.

5.3 Compression and Loss

All knowledge is compression.

A modally differentiated system operating within the constraint-field cannot store, process, or transmit the full complexity of the encounter-patterns it navigates. It must compress: selecting, abstracting, categorising, and stabilising the field into workable patterns that enable orientation and action. Every concept is a compression. Every category is an abstraction. Every model is a simplification.

Compression introduces loss. This is not a defect of the epistemic system. It is the operating condition of any finite system functioning within a field of effectively unlimited constraint-complexity. The loss is structural, not accidental – it follows from the bandwidth limitations of the mediational architecture through which encounter is organised.

3 The foundationalism–coherentism debate has structured anglophone epistemology since at least the mid-twentieth century. Foundationalism (in its classical form) requires a class of self-justifying basic beliefs; coherentism requires only that beliefs support one another in a mutually reinforcing network. Sellars' (1956/1997) critique of the 'myth of the given' undermined classical foundationalism by arguing that no belief is epistemically innocent – all belief involves conceptual classification, and classification presupposes a background framework. MEOW's epistemology inherits Sellars' anti-foundationalism: there are no unmediated epistemic foundations because every stabilisation is structured through mediation. But where Sellars moved toward a broadly coherentist and inferentialist position, MEOW adds an external constraint that pure coherentism lacks: encounter pressure. The constraint-field is not a belief; it is not part of the web of stabilisations. It is what the web must answer to. This is the structural role of constraint-responsiveness in the epistemology: it supplies the external friction that coherentism is frequently accused of lacking.

The consequence is that all knowledge is partial. No stabilisation captures the full constraint-structure of the domain it represents. Scientific models are partial. Perceptual categories are partial. Linguistic descriptions are partial. The partiality is not a temporary limitation to be overcome by better instruments or more careful reasoning. It is a permanent feature of the relationship between finite mediational systems and the constraint-field they inhabit.

This does not make knowledge unreliable. It makes it approximate – and the approximation can be very good. The question is not whether a stabilisation captures reality perfectly (it never does) but whether it captures enough of the relevant constraint-structure to support the navigation, prediction, or coordination for which it is employed. A map is partial. A good map is partial in ways that serve the traveller. A bad map is partial in ways that mislead. The difference is not completeness but alignment with the constraint-patterns that matter for the task at hand.

5.4 Error as Structural

If knowledge is stabilisation under encounter pressure, and all stabilisation involves compression, then error is not an anomaly. It is a structural feature of the epistemic condition.

Error arises from three sources already identified within the framework.

Compression loss. Every stabilisation discards detail. When the discarded detail turns out to be relevant – when the constraint-patterns that were compressed away become operative – the stabilisation fails. This is not carelessness. It is the inevitable cost of compression in a system that cannot represent everything.

Modal divergence. Different agents organise the field differently. Their modal differentiations generate different salience structures, different affordance gradients, different stabilisations of the same constraint-patterns. When agents attempt to coordinate – through language, through institutions – the divergence in their underlying stabilisations produces misalignment. What one agent knows is not what another agent knows, even when both are responding to the same constraint-field, because their mediational architectures compress it differently.

Ontological grammar distortion. The synchronisation protocol through which knowledge is transmitted imposes its own structural biases on the content it encodes (Chapter 7). When these biases are misaligned with the constraint-patterns being described – when substance-grammar forces a processual phenomenon into a noun-shaped category – the resulting stabilisation is systematically distorted. The error is not in the knower's reasoning but in the encoding medium's ontological commitments.

These three sources of error interact. Compression loss reduces the resolution of individual stabilisations. Modal divergence amplifies the discrepancy between agents' stabilisations. Ontological grammar distortion compounds both by shaping the encoding through which stabilisations are formed and transmitted. The result is a system in which error is not merely possible but structurally inevitable – not as a contingent failure of intelligence or diligence but as a consequence of the operating conditions under which knowledge is formed.

Understanding that error is structural rather than accidental has practical consequences. It means that epistemic disagreement need not signal bad faith, irrationality, or moral deficiency. It may signal – and in the Contestables and Fluids regions of the effectiveness gradient, frequently does signal – the predictable divergence of differently mediated stabilisations under compression. The diagnosis is not that one party is wrong and the other right. The diagnosis is that the encoding conditions make convergence difficult or impossible in the domain at issue.

5.5 Knowledge Without Representation

The account developed so far departs from the representational model of knowledge that has dominated Western epistemology since Descartes. On the representational model, knowledge consists of possessing internal representations – beliefs, mental models, conceptual structures – that accurately correspond to external states of affairs. The mind mirrors the world. Knowledge is the fidelity of the mirror.

MEOW replaces this picture. Knowledge is not the possession of internal representations that correspond to an external reality. It is the capacity of a modally differentiated system to maintain effective stabilisations under encounter pressure. Knowing is not depicting reality. It is navigating it – maintaining orientation within a constraint-field that continuously tests, and occasionally vetoes, the stabilisations through which orientation is achieved. The replacement is not merely terminological. It has structural consequences.

First, it dissolves the sceptical problem. Classical scepticism asks: how do you know your representations are accurate? The question presupposes the representational model[4] – it assumes that knowledge requires a match between inner picture and outer reality, and then asks how the match can be verified. Within MEOW, the question does not arise in its traditional form. There is no inner picture to be matched with an outer reality. There is a modally differentiated system operating within a constraint-field, and the system's stabilisations either survive encounter pressure or they do not. The test of knowledge is not correspondence but resilience under constraint.

Second, it reframes objectivity. Objectivity is not the view from nowhere – the God's-eye perspective that sees reality as it is independent of all mediation. There is no such perspective.[5] Objectivity is convergent stability across mediational variation: when differently mediated encounter-systems converge on the same constraint-patterns, that convergence is

4 The strategy of dissolving scepticism by rejecting its presuppositions rather than meeting it on its own terms is shared with several traditions. Dewey argued that the 'spectator theory of knowledge' generates sceptical problems that vanish when knowledge is understood as active inquiry rather than passive reception. Wittgenstein's later philosophy treats sceptical puzzles as symptoms of linguistic confusion rather than genuine epistemic threats. Heidegger's analysis of being-in-the-world dissolves the subject–object gap that scepticism exploits. MEOW's dissolution is closest to Dewey's in spirit: it replaces the representational picture with an account of effective engagement, and the sceptical problem dissolves because the match between inner picture and outer reality was never the right criterion. The difference is that MEOW provides a specific structural account of what replaces representation (stabilisation under constraint-pressure), whereas Dewey's alternative remained more programmatic than formally specified.

5 The phrase 'view from nowhere' originates with Nagel (1986), who used it to characterise the aspiration toward an objective standpoint that transcends any particular perspective. Nagel treated this aspiration as genuine and at least partially achievable. MEOW denies that it is achievable – every encounter is mediated, and no standpoint transcends mediation – but preserves the epistemic work that objectivity is meant to do. The replacement is convergent stability: what is objective is not what is seen from no perspective but what is stable across many perspectives. This is closer to Putnam's (1981) 'internal realism' than to Nagel's aspiration, though MEOW does not share Putnam's conceptual relativism. The convergence is grounded in constraint-pattern invariance, not in the internal coherence of a conceptual scheme.

evidence of structural invariance rather than coincidence of projection. This is the epistemic dimension of the realist anchor developed in Chapter 2 and consolidated in Chapter 9.

Third, it explains why knowledge is cumulative without being foundational. Stabilisations build on other stabilisations. Scientific knowledge accumulates not by laying bricks on a foundation but by constructing an ever more densely interconnected network of stabilisations that mutually constrain and support one another. The network has no bottom. But it has structure, and the structure is constrained by encounter.

5.6 Collective Stabilisation and the Role of Science

Individual stabilisations are fragile. A single agent's compression of the constraint-field is limited by the bandwidth of its mediational architecture, the contingencies of its encounter-history, and the biases of its ontological grammar. Collective stabilisation is more robust.

When multiple agents share stabilisations through the synchronisation protocol – when they compare, critique, and correct one another's compressions – the resulting collective stabilisation is less vulnerable to the idiosyncratic limitations of any individual mediational system. The stabilisation has been tested across multiple modal differentiations. It has survived encounter pressure not just from one perspective but from several.

Science is the most systematic instance of this process. Scientific method does not eliminate mediation. It disciplines stabilisation through structured encounter pressure. Prediction exposes stabilisations to falsification. Experiment subjects them to controlled encounter-conditions. Replication tests them across distinct mediational systems (different laboratories, different instruments, different investigators). Peer review subjects the synchronised description of the stabilisation to the ontological-grammar critique of other agents. Each stage increases the constraint-pressure on the stabilisation. What survives is not truth in the correspondence sense – no stabilisation achieves perfect fidelity to the constraint-field – but truth in the sense that matters: a stabilisation robust enough to support navigation, prediction, and coordination across a wide range of encounter-conditions.

Truth, within this framework, is the long-term survivor of encounter pressure across populations of agents.[6] It is not a property of sentences that

6 This formulation is structurally indebted to Peirce's (1878) convergence theory of truth: 'the opinion which is fated to be ultimately agreed to by all who investigate is what we mean

matches them to states of affairs. It is an achievement of collective stabilisation under constraint.

5.7 Epistemic Confidence and the Effectiveness Gradient

A consequence of this epistemology is that epistemic confidence should scale with the stability of the constraint-pattern being described.

In the Invariants region of the effectiveness gradient (Chapter 7, Figure 7.1) – physical constants, logical relations, well-established causal sequences – stabilisations have survived extensive encounter pressure across radically different mediational architectures. Epistemic confidence is appropriately high. Gravity has been stabilised across human perception, Newtonian mechanics, general relativity, and satellite engineering. The convergence is robust.

In the Contestables region – moral concepts, political categories, aesthetic judgements – stabilisations are less robustly tested, mediational overlap between agents is lower, and ontological grammar distortion is more severe. Epistemic confidence should be correspondingly moderate. Disagreement in these domains is not a sign of epistemic failure. It is the predicted behaviour of finite-bandwidth systems attempting to stabilise constraint-patterns that resist the encoding protocol's discretisation pressure.

In the Fluids region – emotional states, phenomenal qualities, rapidly shifting social dynamics – stabilisations are fragile, convergence is weak, and the protocol's bandwidth is manifestly insufficient. Persistent disagreement and communicative failure are structurally expected.

This scaling of confidence is not scepticism. It does not say that knowledge in the Contestables region is impossible. It says that the operating conditions are harder, the compression loss is greater, and the convergence criteria are more difficult to meet. Knowledge is still achievable. It is just more

---

by the truth'. The chapter's epigraph makes the debt visible. But the framework's version differs from Peirce's in a significant respect. Peirce's convergence is projected to the ideal limit of inquiry – truth is what would be agreed upon if inquiry were pursued indefinitely. MEOW's convergence is structural rather than regulative: it is grounded in the invariance of constraint-patterns across mediational variation, not in a promissory note about the long run. The convergence that grounds truth is not merely projected but empirically evidenced whenever differently mediated encounter-systems arrive at the same constraint-pattern. This makes the realist anchor available now, not only at the idealised end of inquiry.

expensive – it requires more encounter pressure, more collective stabilisation, and more awareness of the encoding biases that shape the result.

5.8 What This Chapter Has Established

Knowledge is stabilisation under encounter pressure. It is not the possession of internal representations that correspond to an external reality. It is the capacity of a modally differentiated system to maintain effective orientation within a constraint-field that continuously tests its stabilisations.

Compression is unavoidable and introduces loss. Error is structural – a consequence of finite bandwidth, modal divergence, and ontological grammar distortion, not a contingent failure of intelligence. The representational model of knowledge is replaced by an account grounded in constraint-sensitive navigation.

Collective stabilisation produces durable knowledge. Science is the most systematic form of collective stabilisation, disciplining individual compressions through structured encounter pressure across multiple mediational architectures. Truth is the long-term survivor of this process.

Epistemic confidence scales with the stability of the constraint-pattern being described, producing a gradient that maps directly onto the effectiveness gradient of Chapter 7. The framework does not abandon realism. It reframes it: reality asserts itself not through perfect transmission but through the persistent failure of bad stabilisations. The world does not reveal itself through correspondence. It reveals itself through refusal.

The question that now follows is: what structural conditions must stabilisations satisfy in order to remain internally coherent? If knowledge is stabilisation under constraint, what constrains the stabilisations themselves? That is the question of logic – and the subject of Chapter 4.

6

Perception

Encounter as Orientation

An affordance is neither an objective property nor a subjective property... It is equally a fact of the environment and a fact of behavior.

— James J Gibson, *The Ecological Approach to Visual Perception*

Nobody sees a flower - really - it is so small it takes time - we haven't time - and to see takes time, like to have a friend takes time.

— Georgia O'Keeffe, *An American Place* (exhibition catalogue)

6.1 Introduction: The Perceptual Question

THE ARCHITECTURE DEVELOPED IN the preceding chapters has established what exists (structured interaction under constraint), what agents are (modally differentiated process-clusters), what logic requires (coherence grammar for stabilisations), and what knowledge is (stabilisation under encounter pressure). What has not yet been addressed is the question that sits between subjecthood and knowledge: how does the constraint-field become structured experience for the systems that inhabit it?

This is the perceptual question. And it is the point at which the framework is most likely to be pulled back toward the representational model it has spent five chapters dismantling.

The representational account of perception is familiar and seductive. The organism receives sensory inputs from the external world. These inputs are processed, interpreted, and assembled into an internal representation – a mental model, a percept, a picture of the environment. Perception is the construction of an inner image from outer data. The organism then acts on the basis of this image.

MEOW rejects this account – not because it is entirely wrong (it captures something about neural processing) but because it reinstalls the substance architecture the framework has dissolved. 'Sensory inputs from the external world' presupposes the mind-world bifurcation. 'Internal representation' presupposes a knowing subject possessing mental objects. 'Outer data' presupposes a noumenal source transmitting signals through a perceptual veil. Every element of the representational story depends on the two-substance picture that Chapter 1 dismantled.

The alternative developed here treats perception not as the construction of inner representations from outer data but as the structured orientation of a modally differentiated system within the constraint-field it inhabits. Perception is not something that happens between a subject and a world. It is something that happens within encounter – the ongoing, mediated, constraint-responsive engagement of a self-maintaining system with the field in which it operates.

6.2 Affordances as Encounter-Constituted Gradients

The concept that does the central work in this chapter is affordance, drawn from Gibson's ecological psychology but reframed within MEOW's architecture.

Gibson proposed that organisms do not perceive a neutral world of objects and properties and then compute what to do with them. They perceive affordances[1] – possibilities for action that the environment offers relative to the organism's capacities. A flat, rigid, knee-high surface affords sitting for a human. A narrow ledge affords perching for a pigeon. The affordance is not a property of the surface alone, nor a property of the organism alone. It is a relational fact – a feature of the organism-environment system.

MEOW inherits this relational insight and sharpens it in two ways.

First, affordances are constituted by encounter rather than belonging to the subjectless field as such. This distinction, flagged in the revised table of contents and developed here, matters because it determines the ontological status of affordances within the framework. The constraint-field exists independently of encounter – that is MEOW's realist commitment. But 'independently of encounter' does not mean 'independently specifiable without mediation'. It means the constraint-patterns are not dependent on an organism's endorsement for their persistence. Affordances, however, do not exist independently of encounter. An affordance is a viable trajectory through the constraint-field relative to a particular modally differentiated system. Without the system, there is constraint topology – structure, invariance, resistance. With the system, there are affordances – navigable gradients, actionable possibilities, salient pathways. Affordances are encounter-constituted. They arise when a self-maintaining system differentiates the field into what supports and what obstructs its persistence.

Second, affordances are gradients, not binary switches. A surface does not simply afford or not-afford walking. It affords walking more or less easily, depending on slope, friction, stability, and the organism's bodily capacities.

1 Gibson's concept of affordances has been developed in several competing directions since its original formulation in *The Ecological Approach to Visual Perception* (1979). Turvey (1992) formalised affordances as dispositions of the animal–environment system. Chemero (2003, 2009) argued that affordances are relations between abilities and features of the environment, rejecting the dispositional analysis. Reed (1996) developed an 'ecological ontology' grounded in affordances as resources. MEOW's treatment – affordances as encounter-constituted compatibility gradients – is closest to Chemero's relational reading but goes further: affordances are not merely relational facts about an organism and its environment; they are constituted by the encounter-event itself and do not exist as affordances prior to it. The constraint topology exists independently; its articulation as navigable gradients does not. This distinction (constraint topology vs. affordance gradient) is MEOW's specific contribution to the affordance literature.

An environment does not simply afford or not-afford shelter. It affords degrees of shelter relative to the threat being sheltered from, the organism's thermal requirements, and the available materials. Affordances are compatibility gradients between the organism's organisational profile and the constraint topology of the field. They are continuous, context-sensitive, and multiply realisable. Constraint topology is encounter-independent structure; affordance is that structure rendered navigable relative to a particular organisational profile.

This framing avoids a problem that Gibson's own account never fully resolved: the question of whether affordances are objective features of the environment or subjective projections of the organism. Within MEOW, the question dissolves. Affordances are neither objective (they do not exist independently of the encountering system) nor subjective (they are not projected by a mind onto a neutral world). They are relational – constituted by the encounter between a particular mediational architecture and the constraint-patterns of the field. They are real, but their reality is encounter-relative, not encounter-independent.

6.3 Meaning as Salience-Weighted Affordance Structure

It is here that meaning is forced into view – not as a property of language (that comes in Chapter 7) but as a structural feature of perception itself.

Meaning, within MEOW, is salience-weighted affordance structure relative to organismal persistence.[2] For humans, persistence conditions include not only metabolic continuity but social continuity, role-stability, and symbolic commitments – which is why human meaning-structures are richer

2 The claim that meaning is pre-linguistic and grounded in organismal engagement with the environment has a rich phenomenological ancestry. Merleau-Ponty's (1945/1962) analysis of the body-subject's motor intentionality locates meaning in the lived body's practical engagement with its surroundings, prior to any reflective or linguistic act. Heidegger's (1927/1962) account of worldhood grounds significance in the equipmental totality within which Dasein operates: the hammer is meaningful as ready-to-hand before it is thematised as an object of contemplation. MEOW's formulation shares the structural commitment of both – meaning arises in practical engagement, not in linguistic labelling – but grounds it in the more austere vocabulary of encounter-organisation rather than in phenomenological description. The advantage is precision; the cost is the loss of Merleau-Ponty's and Heidegger's descriptive richness, which is acknowledged here rather than defended as a trade-off worth making.

and more layered than those of organisms whose persistence is purely biological. An encounter is meaningful when the affordance gradients it discloses are weighted by the self-maintaining system's persistence conditions – when some affordances are more urgent, more relevant, or more consequential than others relative to what the organism needs to continue operating.

A predator approaching is meaningful because it weights the affordance gradient sharply: escape routes become salient, hiding places acquire urgency, irrelevant features of the environment recede. A ripe fruit is meaningful because it weights the affordance gradient toward acquisition: the colour becomes salient, the path to it becomes navigable, competing demands are deprioritised. In each case, meaning is not added to perception by an act of interpretation. It is constituted by the asymmetric organisation of the affordance structure relative to the organism's persistence conditions.

This account has two consequences that matter for the rest of the monograph.

First, meaning is pre-linguistic. It arises in the encounter between a self-maintaining system and the constraint-field before any linguistic encoding occurs. Organisms without language – animals, pre-linguistic infants – inhabit meaningful environments. Their encounters are organised by salience-weighted affordance gradients that orient action without the mediation of words. Language, when it arrives (Chapter 7), does not create meaning. It compresses and synchronises pre-existing meaning-structures across agents. This ordering – meaning first, language second – prevents the framework from collapsing into the view that meaning is a linguistic artefact, which would be both empirically implausible and architecturally incoherent within MEOW.

Second, meaning is encounter-relative without being arbitrary. The affordance gradients that constitute meaning are constrained by the field's structure. An organism cannot mean the predator away. It cannot mean the fruit into existence. The salience-weighting is the organism's contribution; the constraint-patterns that are weighted are the field's. Meaning arises from the interaction between the two. This is how MEOW avoids idealism (meaning is not projected onto a blank world) while preserving phenomenology (meaning is experientially real within encounter).

6.4 Perception Without Representation

The account of perception that emerges from these commitments is non-representational. It does not deny that neural systems process information, construct models, or generate predictions. It denies that these processes are best understood as the construction of inner pictures that correspond to an outer reality.

Perception, within this framework, is the ongoing structured orientation of a modally differentiated system within its affordance gradients. The organism does not build a model of the world and then consult it. It continuously adjusts its orientation within the constraint-field in response to the affordance gradients that its modal differentiation discloses. Perception is not a product (a completed representation) but a continuous process (an ongoing adjustment of orientation).

The distinction has consequences. If perception is representation, then perceptual error is misrepresentation – the inner picture fails to match the outer reality. If perception is orientation, then perceptual error is misalignment – the organism's stabilisation of the affordance gradient diverges from the constraint-patterns that actually obtain. Both accounts predict the same phenomenology (the organism gets things wrong), but they differ in their metaphysical commitments. The representational account requires the two-substance architecture (inner model, outer world). The orientation account requires only the encounter-architecture already established: a modally differentiated system navigating a constraint-field whose affordance gradients are partially but not perfectly disclosed.

Predictive processing – the influential framework in computational neuroscience that treats perception as the brain's best prediction of sensory input, continuously updated by prediction error – is compatible with MEOW's account, provided its representational language is reinterpreted.[3]

3 The relationship between predictive processing and anti-representationalism is contested within the literature itself. Clark (2013, 2016) generally retains representational language while emphasising action-oriented prediction. Hohwy (2013) defends a more robustly representational reading in which the brain literally models the causal structure of the world. Radical enactivist critiques (Hutto & Myin, 2013) argue that predictive processing is incompatible with genuine anti-representationalism and that its use of terms like 'model' and 'prediction error' smuggles in the very picture it claims to supersede. MEOW's position is a fourth option: it accepts the computational architecture (hierarchical prediction, error-correction, active inference) while denying the ontological interpretation (inner

Within MEOW, prediction is not the construction of an inner model that is compared with outer data. It is the anticipatory stabilisation of affordance gradients – the system's ongoing attempt to maintain coherent orientation by generating expectations about the constraint-field and revising them when encounter-pressure disconfirms them. Prediction error is not a mismatch between model and world. It is a signal that the system's current stabilisation of the affordance gradient needs revision. The computational machinery is the same. The metaphysical interpretation is different.

6.5 Cross-Modal Integration and the Unity of Perceptual Encounter

A modally differentiated system does not encounter the constraint-field through a single sensory channel. Vision, audition, proprioception, haptic contact, olfaction, interoception – each is a distinct mediational pathway through which the field is structured into affordance gradients. Perception, in the ordinary sense, involves the integration of these pathways into a unified encounter-organisation.

This integration is not the assembly of separate inputs into a combined representation – that framing reinstalls the representational model. It is the co-stabilisation of affordance gradients across mediational pathways. The visual disclosure of a surface's slope, the haptic disclosure of its texture, the proprioceptive disclosure of the body's balance – these are not separate data streams that are subsequently combined. They are concurrent aspects of a single encounter-organisation in which the system's modal differentiation is structured by multiple mediational pathways simultaneously.

The unity of perceptual experience – the fact that we encounter a coherent environment rather than a cacophony of separate sensory signals – is therefore not a puzzle to be explained by positing a binding mechanism that glues separate representations together. It is the default condition of modal differentiation: the self-maintaining system is a single organisational centre whose modal differentiation structures the constraint-field holistically. The 'binding problem' of neuroscience arises because the representational frame-

model corresponding to outer world). This is possible because the computational story can be retold in terms of anticipatory stabilisation and encounter-pressure without loss of explanatory power. The question is not whether the brain predicts. It is whether 'prediction' requires a representational metaphysics.

work fragments perception into separate channels that then require reunification. MEOW does not fragment perception in the first place.[4] The unity is the default condition. The fragmentation is an artefact of the analytical approach, not a feature of the encounter.

6.6 Perceptual Variation and Shared Constraint

Different organisms perceive differently. This is an empirical commonplace. Bees perceive ultraviolet. Bats perceive through echolocation. Sharks perceive electrical fields. Humans perceive through a particular configuration of trichromatic vision, binaural audition, and bodily proprioception. Each mediational architecture discloses a different affordance-profile of the same constraint-field.

The variation is real and irreducible. There is no 'view from nowhere' that combines all possible perceptual encounters into a single master-perception. Each organism's modal differentiation generates its own structured orientation within the field. The worlds they inhabit are phenomenologically different.

And yet, the constraint-field is shared. The spectral regularities that constrain the bee's ultraviolet perception and the human's trichromatic perception are the same underlying constraint-patterns, differently disclosed. The acoustic regularities that constrain bat echolocation and human audition are differently mediated encounters with a shared constraint-structure. Cross-species convergence on the same environmental features – both bees and humans avoid obstacles, navigate toward food sources, respond to predator cues – is evidence of shared constraint, not coincidence of projection.

4 The binding problem – how the brain integrates information processed in anatomically distinct regions into unified conscious experience – has generated a substantial neuroscientific literature (Treisman, 1996; Singer & Gray, 1995; Roskies, 1999). Proposed mechanisms include temporal synchrony (neural oscillations binding distributed representations), feature integration through attention, and re-entrant processing. MEOW's claim is not that these neuroscientific findings are wrong but that the problem's framing is an artefact of the representational assumption. If perception is the construction of separate feature-representations that must be bound, binding is a genuine puzzle. If perception is the holistic orientation of a modally differentiated system, the unity is the starting condition and the analytical decomposition into separate features is the secondary operation. The neuroscientific mechanisms identified as 'binding' mechanisms may be better understood as mechanisms that maintain coherent orientation across mediational pathways, rather than mechanisms that glue separate representations together.

This is the perceptual dimension of the realist anchor developed in Chapter 2 and consolidated in Chapter 9. Objectivity in perception is not the achievement of an unmediated view. It is the convergence of differently mediated encounters on shared constraint-patterns. When organisms with radically different mediational architectures respond to the same environmental features, the convergence is evidence that the features are genuinely there – not as objects in a substance-ontological sense, but as invariant constraint-patterns within the field that are disclosed, differently but convergently, by distinct modal differentiations.

6.7 Perception, Language, and the Encoding of Affordance

The relationship between perception and language is a source of persistent confusion in philosophy, and MEOW's architecture clarifies it by enforcing an ordering: perception is prior; language is subsequent.

Perception structures encounter into affordance gradients. Language compresses those affordance-structures into transmissible form. The two are not the same process. Perception is the ongoing orientation of a modally differentiated system within the constraint-field. Language is a second-order operation that takes the products of perceptual orientation – stabilised affordance-patterns – and encodes them for synchronisation across agents.

The encoding inevitably transforms what it encodes. Chapter 7 develops this claim in detail: the synchronisation protocol imposes ontological grammar on the affordance-structures it transmits, discretising gradients, nominalising processes, and installing substance-ontological commitments. But the transformation operates on material that is already meaningful – on affordance-structures that are already weighted, already organised, already perceptually constituted. Language does not create meaning from meaningless data. It compresses and distorts pre-linguistic meaning into a format that enables coordination at the cost of fidelity.

This ordering prevents two errors. First, it prevents the view that perception is 'raw' and meaning is 'added' by language – a view that would reinstall the representational model (perception supplies data; language supplies interpretation). Perception is already meaningful. Language operates on that meaning. Second, it prevents the view that perception is 'theory-laden' in a way that makes it entirely dependent on linguistic categories[5] – a view that would collapse perception into language and make pre-linguistic perception unintelligible. Perception is shaped by mediation (biological, cognitive, cultural), but it is not shaped by language specifically except insofar as language is one mediational layer among several. An organism without language still perceives meaningfully. The affordance gradients are there before the words arrive.

6.8 Engagement with Adjacent Traditions

The account developed here draws on and departs from three adjacent traditions in philosophy and cognitive science. The departures are noted here briefly; specialists may consult Appendix B for a fuller genealogical treatment.

Gibson's ecological psychology provides the concept of affordances and the anti-representational orientation that MEOW inherits. The departure is twofold. Gibson treated affordances as objective features of the environment – as genuinely 'out there' independent of the organism. MEOW treats them as encounter-constituted: relational, arising in the organism-field interaction, not locatable on either side of the relation independently. Additionally, Gibson's account carries residual teleological implications – the environment 'invites' or 'calls for' action. MEOW strips this out. Affordances are compatibility gradients along which metastable systems can move. They are not invitations. The field does not beckon.

Enactivism (Varela, Thompson, Maturana) shares MEOW's anti-representational commitment and its emphasis on organism-environment coupling. The departure concerns scope. Enactivism typically restricts its account to biological systems – to living organisms whose autonomy and

5 The theory-ladenness of perception, argued most forcefully by Hanson (1958) and developed by Kuhn (1962/2012) and the Churchlands, holds that what we perceive is shaped by the theoretical commitments we bring to the encounter. The thesis exists in strong and weak forms. The strong form – that perception is constitutively dependent on theory or linguistic categories – is what MEOW rejects. It would collapse perception into a species of linguistic processing, making pre-linguistic perception unintelligible. The weak form – that theoretical commitments and linguistic categories are among the mediational factors that shape encounter-organisation – is entirely compatible with MEOW's framework. Language is one mediational layer among several. It shapes encounter at the linguistic–cultural level of mediation (Chapter 1, §1.4) without being the sole or foundational source of perceptual structure. The ordering – perception first, language second – is what preserves this distinction.

sense-making capacities are grounded in autopoiesis. MEOW's formulation is substrate-neutral: the organisational profile rather than the biological substrate determines whether modal differentiation obtains. This does not mean MEOW denies the importance of biology. It means the framework does not make biology a definitional requirement for encounter-organisation, which preserves its range across non-biological systems without committing to claims about their perceptual capacities.

Predictive processing (Clark, Friston, Hohwy) offers a computational framework for understanding perception as prediction under uncertainty. MEOW can reinterpret the computational machinery of predictive processing but rejects its metaphysical commitments. Predictive processing, in some formulations, retains representational language: the brain builds a 'generative model' of the world and minimises 'prediction error' against sensory data. Within MEOW, these are anticipatory stabilisations of affordance gradients and encounter-pressure signals indicating misalignment, respectively. The computational story is the same. The ontological framing – inner model versus outer world – is replaced by encounter-orientation within a constraint-field.

6.9 What This Chapter Has Established

Perception is the structured orientation of a modally differentiated system within the affordance gradients of the constraint-field it inhabits. It is not the construction of internal representations from external data.

Affordances are organism-field compatibility gradients constituted by encounter. They arise when a self-maintaining system differentiates the field into what supports and what obstructs its persistence. They are relational, continuous, and context-sensitive – neither objective properties of the environment nor subjective projections of the organism.

Meaning is salience-weighted affordance structure relative to organismal persistence. It is pre-linguistic: it arises in encounter before any linguistic encoding occurs. Language compresses and synchronises pre-existing meaning-structures across agents, at the cost of fidelity and with the imposition of ontological grammar.

Perceptual variation across organisms is real and irreducible. Perceptual objectivity is not the achievement of an unmediated view but the convergence of differently mediated encounters on shared constraint-patterns. This convergence is the perceptual dimension of the realist anchor.

The architecture of Part II is now complete. Encounter-events (Chapter 1), constraint and ontology (Chapter 2), subjecthood (Chapter 3), logic (Chapter 4), epistemology (Chapter 5), and perception (this chapter) constitute the core engine of the framework. Part III extends the engine into language (Chapter 7) and social ontology (Chapter 8). Part IV addresses the realist defence, demonstrates the framework's explanatory purchase, and identifies its limits.

7

Language

Synchronisation, Ontological Grammar, and Structural Limits

The limits of my language mean the limits of my world.
Die Grenzen meiner Sprache bedeuten die Grenzen meiner Welt

— Ludwig Wittgenstein, *Tractatus Logico-Philosophicus*

Human speech is like a cracked kettle on which we tap crude rhythms for bears to dance to, while we long to make music that will melt the stars.
La parole humaine est comme un chaudron fêlé où nous battons des mélodies à faire danser les ours, quand on voudrait attendrir les étoiles

— Gustave Flaubert, *Madame Bovary*

7.1 Introduction: Language as Infrastructure

Language is the most conspicuous feature of human cognitive life. It is also the most routinely mischaracterised. Philosophy has treated it as a mirror of thought, a system of names for things, a medium of representation, and a vehicle for truth. Each of these framings captures something, but each installs assumptions that distort the phenomenon.

Within the architecture of this monograph, language occupies a particular structural position. It is not a window onto reality. It is not a container for thought. It is infrastructure – a coordination mechanism through which encounter-patterns are compressed into transmissible form and approximately reconstructed across distinct mediational systems. Language, on this view, is a finite-bandwidth synchronisation protocol: a means of aligning predictive stabilisations across multiple organisms operating within overlapping but non-identical encounter-regimes.

This framing has three consequences that the present chapter develops.

First, it explains what language does – not naming or representing, but coordinating. Language enables organisms with distinct mediational architectures to operate within shared constraint-environments by synchronising their stabilisations. When this synchronisation succeeds, we get communication, cooperation, and the extraordinary apparatus of human social life. When it fails – and it fails systematically – we get misunderstanding, conceptual impasse, and the phenomena diagnosed by the *Language Insufficiency Hypothesis*.

Second, it reveals what language smuggles in. The grammatical structures of natural language are not ontologically neutral. They impose commitments – about what kinds of entities exist, what kinds of relations obtain, and how persistence, change, and identity are organised – before any philosophical reflection begins. This is the ontological grammar thesis, and it explains why the substance ontology rejected in Chapter 2 has proven so resistant to displacement.

Third, it identifies where language breaks down. The synchronisation protocol has bandwidth limitations that are structural rather than incidental. In domains of high fluidity, weak convergence, or radical mediational asymmetry, language systematically fails to synchronise stabilisations across agents. This is not a defect to be remedied by better vocabulary. It is a consequence of the protocol's operating conditions. The Language Insufficiency Hypothesis formalises this structural limitation.

These three threads – synchronisation, ontological grammar, and structural limits – are not independent topics housed under a single chapter heading. They are aspects of a single phenomenon. Language synchronises, but only within the constraints of its grammatical structure. That structure shapes what can be synchronised and how. And the shaping has limits beyond which the protocol ceases to function reliably. The chapter develops this integrated account.

7.2 Language as Synchronisation Protocol

Human beings do not share a single mediational architecture. Each organism encounters reality through its own biological, cognitive, and experiential organisation. Perceptual systems differ in sensitivity and bandwidth. Cognitive histories differ in training, association, and expectation. Experiential trajectories – the sequences of encounter-events through which an organism has passed – are unique to each individual.

And yet human beings coordinate. They share projects, build institutions, transmit knowledge across generations, and conduct arguments about metaphysics. This coordination requires a mechanism for aligning stabilisations across non-identical encounter-systems.

Language is that mechanism. If encounter-events are primary and constraint-patterns stabilise across mediation, then language operates as a second-order stabilisation mechanism layered atop this architecture – compressing first-order encounter-patterns into transmissible form.

The process can be described schematically. An organism encounters a constraint-pattern within its own mediational architecture. It stabilises that encounter – categorises it, associates it with prior encounters, integrates it into a predictive model. It then compresses that stabilisation into a linguistic form: a word, a phrase, a sentence, a narrative. The linguistic form is transmitted – spoken, written, signed – to another organism, which decompresses it within its own mediational architecture, reconstructing an approximate stabilisation.

The reconstruction is never identical to the original. It cannot be. The receiving organism's mediational architecture differs from the sender's.

What is reconstructed is an approximation – a stabilisation that is structurally similar enough to enable coordination but never a perfect replica of the original encounter-pattern. Communication is not duplication. It is approximate synchronisation under constraint.

This framing reorients several traditional problems in philosophy of language.

But first, a more fundamental reorientation must be noted. The account of language as synchronisation protocol stands in direct tension with the convergence assumption that underwrites orthodox communication theory. Shannon-Weaver, Jakobson, Gricean pragmatics, and their descendants all treat convergence as the default condition of communication.[1] The models assume that transmission succeeds unless something goes wrong – noise in the channel, ambiguity in the encoding, violation of cooperative principle. Divergence is aberration. Convergence is baseline. The entire apparatus is built around identifying and correcting departures from successful transmission.

This assumption is not empirically substantiated. It is an artefact of the models' architecture. Later developments in pragmatics and information theory – relevance theory, computational models of communication, Bayesian pragmatics – address some of the limitations of the original Shannon-Weaver framework. The point here concerns a deeper assumption that persists across the lineage: the treatment of convergence as the default condition of communication rather than a structurally constrained achievement.

1 The characterisation of Shannon–Weaver as assuming convergence requires qualification. Shannon's (1948) mathematical theory of communication is strictly concerned with signal transmission fidelity – with whether the received signal reproduces the sent signal – and is explicitly agnostic about meaning. The convergence assumption identified here is not present in Shannon's formalism but in the philosophical appropriation of that formalism: the widespread tendency to model communication as successful signal transmission plus interpretation, treating divergence as noise rather than as a structural feature of the system. Grice's (1989) cooperative principle and its maxims similarly presuppose a baseline of shared intention and common ground. Later developments – Sperber and Wilson's (1986) relevance theory, Bayesian pragmatics – relax some of these assumptions but retain the orientation toward convergence as the norm from which departures are explained. The main text's claim targets this shared orientation, not the mathematical details of any particular model.

Shannon-Weaver abstracts away from mediational architecture entirely, treating sender and receiver as functionally equivalent terminals. The model has no place for the possibility that decompression at the receiving end produces a fundamentally different stabilisation than the one compressed at the sending end, because it does not model the mediational systems doing the compressing and decompressing. Gricean pragmatics presupposes that speaker and hearer share sufficient common ground for implicature to function – but 'shared common ground' is itself a stabilisation that obtains only under specific conditions, and it degrades along the same gradient as everything else. These models treat the conditions for convergence as given. MEOW treats them as achievements – structurally constrained achievements that succeed only when constraint-patterns are stable, mediational overlap is sufficient, and the domain admits of discrete restabilisation.

The most prominent defence of the convergence assumption – familiar in popular cognitive science and some pragmatic traditions, associated most visibly with Pinker – argues that language's long history of successful coordination is itself evidence of its adequacy.[2] But the evidence marshalled is drawn overwhelmingly from domains where the protocol does succeed: scientific communication, technical instruction, object-reference, spatial navigation, transactional exchange. These are Invariants-region phenomena – stable constraint-patterns, high discretisation tolerance, strong mediational overlap. The success is real. The inference from it to a general claim about language's adequacy is not. It is equivalent to testing a bridge with bicycles and concluding it can carry freight trains. Language's performance in domains of stable, well-discretised constraint tells us nothing about its performance in the *Contestables* and *Fluids* regions, where the operating

2 Pinker's (2007) *The Stuff of Thought* argues that language is a window into human nature, broadly adequate to the expressive demands placed upon it and shaped by evolutionary pressures toward communicative success. The present chapter does not dispute the evolutionary account – it is entirely plausible that the synchronisation protocol has been shaped by selection pressure toward coordination efficiency. What it disputes is the inference from evolutionary adequacy to general adequacy. A protocol shaped by selection for coordination in environments of shared physical constraint (predator avoidance, resource location, tool use) may be superbly adequate for those domains and structurally inadequate for domains the selection pressure never targeted – precisely the *Contestables* and *Fluids* regions of the effectiveness gradient. The evolutionary argument is a special case of the sampling bias identified in the main text: testing the protocol where it was designed to succeed and generalising to domains where it was not.

conditions are structurally different and the evidence of genuine convergence is conspicuously thinner. What exists in those regions is evidence of apparent convergence – speakers using the same words, reaching nominal agreement, and discovering downstream that the stabilisations they reconstructed were not the same.

The inversion matters. If convergence is the default, then communicative failure is a bug – something went wrong that can be identified and fixed. If convergence is the achievement, then communicative failure in high-complexity domains is the predicted behaviour of the system operating beyond its structural parameters. The implications for political discourse, institutional reasoning, and cross-cultural communication are immediate: what is experienced as bad faith, irrationality, or wilful misunderstanding may in many cases be the protocol functioning exactly as its structural constraints predict. This reframing is developed further in Sections 7.7 through 7.9, and its institutional consequences are the subject of Chapter 8.

Meaning is not a property of words or sentences. It is a coordination achievement – the degree to which linguistic compression and decompression succeed in aligning stabilisations across distinct mediational systems. Meaning is therefore not fixed but negotiated: it depends on the overlap between the mediational architectures of sender and receiver, the stability of the constraint-patterns being communicated, and the resolution of the linguistic form used to compress them.

Reference is not a word-world relation. It is a convergence indicator – a linguistic device that succeeds when it directs two or more mediational systems toward the same constraint-pattern.[3] 'The boulder' works as a referring expression not because it hooks onto a substance in the world but

3 The reframing of reference as convergence indicator rather than word–world correspondence departs from both the causal theory of reference (Kripke, 1980; Putnam, 1975) and the descriptivist tradition (Frege, Russell). Causal theories ground reference in a historical chain connecting a name to an initial 'baptism' of the referent. Descriptivist theories ground it in the satisfaction of descriptive conditions. MEOW replaces both with a functional account: reference succeeds when it produces convergent stabilisation across mediational systems. 'The boulder' refers not because it causally connects to a substance or satisfies a description, but because it reliably synchronises encounter-stabilisations among agents operating at similar scales. The account is deflationary about the metaphysics of reference but not about its practical importance: convergent reference is the condition for linguistic coordination, and its failure in the Contestables region is one source of the gradient's degradation.

because it reliably synchronises encounter-stabilisations across organisms operating at similar scales.

Truth is not correspondence between sentences and states of affairs. Within this framework, truth is the success condition for synchronisation under constraint:[4] a statement is true to the extent that the stabilisation it transmits converges with the constraint-patterns accessible to the receiving system, where the constraint-pattern itself is not up for negotiation. This is not deflationism and not pragmatism, though it shares features with both. It is a structural account of truth grounded in the architecture of encounter.

7.3 What Grammar Smuggles In: The Ontological Grammar Thesis

Language does not merely transmit stabilisations. It shapes them. And the shaping is not optional.

Every natural language imposes a grammatical structure on the encounter-patterns it encodes. This structure determines what kinds of stabilisations can be transmitted, how they are organised, and what ontological commitments are embedded in the act of transmission. These commitments are not philosophical positions adopted by speakers. They are architectural features of the protocol itself – built into the grammar before any speaker begins to philosophise.

The claim requires precision. It is not that language determines thought. It is that grammatical structure scaffolds ontological reflexes – habitual patterns of categorisation that feel inevitable because they are grammatically enforced.[5]

---

4 The account of truth offered here should be distinguished from three neighbours it may be confused with. Deflationism (Horwich, 1990; Field, 2001) holds that 'true' is a logical device without substantial metaphysical content; MEOW agrees that truth is not a heavyweight metaphysical property but disagrees that it is merely logical – it is a structural achievement of convergent synchronisation. Pragmatism (James, 1907; Dewey, 1938) treats truth as what works in practice; MEOW's account is structurally adjacent but specifies what 'working' consists of: surviving encounter pressure across mediational variation. Habermas's (1984) discourse theory grounds truth in the conditions of rational consensus; MEOW shares the emphasis on inter-agent convergence but locates the convergence in constraint-pattern stability rather than in rational agreement. The constraint-field is not a conversation partner. It is the structure that conversations must answer to.

5 The ontological grammar thesis occupies territory between several established positions in the philosophy of language. Whorf's (1956) linguistic relativity, discussed in §7.4, claims

Indo-European languages – the languages in which the Western philosophical tradition has been conducted – exhibit a characteristic grammatical architecture. The basic unit of expression is the subject-predicate sentence: a noun (naming a substance) bears a verb (describing an action or state) or an adjective (attributing a property). 'The rock falls.' 'The mind perceives.' 'The electron orbits.' Each sentence presupposes an entity that exists prior to and independent of what it does. The grammar installs substance ontology at the level of syntax.

This installation has several consequences.

Objects become default ontology. Because grammar requires noun-subjects, speakers are habituated to a world composed of discrete, bounded, persistent things. Processes must be nominalised to enter discourse: 'running' becomes 'a run', 'thinking' becomes 'a thought', 'interacting' becomes 'an interaction'. The verb is converted into a noun, and the process is reified into an object. This is not a philosophical error committed by careless thinkers. It is a structural feature of the encoding protocol.

Relations become secondary. In subject-predicate grammar, relations are expressed as predicates borne by subjects: 'A is taller than B', 'the mind perceives the world'. The relata come first; the relation is added. This grammatical order mirrors – and reinforces – the substance-ontological assumption that relata are prior and relations derivative. MEOW's inversion of this priority (relations are primitive, relata are derivative) is therefore not merely counterintuitive. It is counter-grammatical. It works against the encoding structure of the language in which it must be articulated.

Change becomes exceptional. In noun-dominant grammar, the default state of an entity is persistence. Change is something that happens to a

---

that grammatical categories shape habitual thought. Quine's (1969) ontological relativity argues that what counts as an object depends on the background theory. Heidegger's observations about the 'house of Being' suggest that language does not merely express but constitutes the horizon of intelligibility. Wittgenstein's (1953) 'language games' locate meaning in use, which implies that the structural features of the game shape what can be meaningfully said within it. The ontological grammar thesis draws on all of these without reducing to any. Its specific claim is narrower and more mechanistic than Heidegger's or Wittgenstein's: it identifies subject-predicate syntax as the structural feature that installs substance-ontological commitments, and it traces the consequences of this installation across specific philosophical problems (the no-relata demand, the mind–world bifurcation, the hard problem). The thesis is not that language constitutes reality (idealism) but that grammar shapes the default categories through which reality is stabilised and transmitted.

substance – an event that modifies a pre-existing thing: 'The ice melted.' 'The empire fell.' 'The patient recovered.' Each sentence presupposes a stable entity that undergoes alteration. The durative priority proposed in Chapter 2 – that stability is slow interaction and identity is stabilised duration – reverses this grammatical default. In the grammar, being precedes becoming. In the ontology, becoming precedes being. The discomfort many readers will feel with this reversal is partly philosophical and partly grammatical. The grammar makes the substance-first picture feel like common sense.

The mind-world bifurcation is grammatically generated. 'The mind perceives the world' is a perfectly grammatical sentence. It installs two noun-subjects – mind, world – and a verb-relation – perceives – between them. The sentence presupposes the very architecture MEOW rejects: two independently existing substances entering into epistemic contact. A language that did not require noun-subjects, or that treated relations as grammatically primary, would not generate the same metaphysical picture. The realism-idealism dispute, at least in part, is an artefact of the grammar in which it has been conducted.

7.4 Ontological Grammar Is Not Linguistic Relativism

A clarification is needed here, because the ontological grammar thesis will be confused with the Sapir-Whorf hypothesis if the distinction is not drawn explicitly.

The Sapir-Whorf hypothesis, in its strong form, claims that the structure of a language determines the thought patterns of its speakers – that linguistic categories constrain cognition in ways that cannot be transcended. In its weaker form, it claims that language influences habitual thought without strictly determining it. Both versions operate at the level of cognition: the question is whether speakers of different languages think differently because of their linguistic structures.

The ontological grammar thesis is a different kind of claim. It is not primarily about cognition. It is about the ontological commitments embedded in the encoding protocol itself. The claim is not that speakers of Indo-European languages cannot think in durative or relational terms – manifestly they can, as the existence of process philosophy, Buddhist metaphysics, and this monograph attests. The claim is that the grammatical structure of these languages makes substance-ontological framings cognitively cheaper and relationally-ontological framings cognitively more expensive. The grammar does not imprison thought. It taxes alternatives. The principle of least effort does the rest.

This is a structural observation about the economics of articulation, not a deterministic claim about the limits of cognition. Speakers can and do think against their grammars. But doing so requires effort, and the default – the path of least resistance – flows toward whatever ontological picture the grammar rewards with fluency. In Indo-European languages, that picture is substance ontology. The ontological grammar thesis explains why substance thinking persists even after its philosophical inadequacies have been exposed. It persists because the grammar makes it easy and the alternatives hard.

7.5 Durative Grammar and the Substance Alternative

If noun-dominant grammar scaffolds substance ontology, what would a grammar look like that scaffolded the durative, interaction-first ontology proposed by this monograph?

The question is not merely hypothetical. Several natural languages weight their grammatical structures differently from the Indo-European pattern.[6] Many Indigenous Australian languages foreground relational and directional structures over noun-based object grammar. Mandarin Chinese deploys aspect-marking rather than tense-marking in ways that treat events as primary and temporal framing as contextual (Evans & Levinson, 2009). None of these examples should be romanticised or treated as evidence that speakers of these languages have 'better' metaphysics. They should be treated as evidence that the relative prominence of nominal versus verbal or relational structures varies across languages, and that the substance-ontological

6 The cross-linguistic evidence for grammatical variation in ontological scaffolding is extensive but requires careful handling. Lucy (1992) provides the most rigorous case study of how grammatical number-marking in Yucatec Maya correlates with differences in categorisation behaviour relative to English. Boroditsky (2001) documents effects of grammatical metaphor on temporal reasoning in Mandarin versus English. Evans and Levinson (2009), cited in the main text, argue that putative language universals are far fewer than the Chomskyan tradition assumes. The present chapter does not claim that speakers of non-Indo-European languages have access to better metaphysics. It claims that the variation demonstrates that the substance-ontological scaffolding of Indo-European grammar is a feature of a particular grammatical architecture rather than a universal feature of language. The cross-linguistic evidence is used diagnostically, not prescriptively.

defaults of Indo-European grammar are features of a particular grammatical architecture rather than universal features of language as such.

Within the framework of this monograph, the relevant contrast is between:

Noun-dominant grammar, in which being precedes becoming, entities are primitive, and relations are predicated of pre-existing subjects. This scaffolds substance ontology. It makes object-first thinking fluent and relational thinking cumbersome.

Durative or process-dominant grammar, in which becoming precedes being, interactions are primitive, and entities emerge as stable configurations within ongoing process. This scaffolds relational and process ontologies. It makes interaction-first thinking more natural but object-individuation more effortful.

Neither grammar is 'correct'. Both are encoding protocols with characteristic compression profiles. The point is that the choice of grammar has ontological consequences that are rarely examined because the grammar itself is transparent to its users. Speakers do not notice the ontological commitments of their grammar any more than fish notice water. The commitments become visible only when an alternative grammar – or an alternative ontology – makes them salient.

Bergson recognised something structurally similar (Bergson, 1910). His critique of the intellect's tendency to spatialise duration – to represent continuous process as a series of discrete static states – identifies the same distortion that the ontological grammar thesis locates in noun-dominant syntax. Where Bergson attributed the distortion to the structure of intellect itself and proposed intuition as the remedy, the present account locates it more precisely in grammatical architecture and proposes not the abandonment of conceptual thought but awareness of the encoding biases it carries. The diagnosis converges. The remedy diverges. Bergson's intuitionism – the claim that only non-conceptual, non-linguistic awareness can grasp duration – is both methodologically unusable and in tension with the commitment to structural articulation that this monograph embodies. The ontological grammar thesis preserves Bergson's diagnostic insight whilst discarding his epistemological prescription.

7.6 What Grammar Does to Philosophy

The consequences of ontological grammar for philosophical practice are worth making explicit, because they bear directly on the reception of the framework developed in this monograph.

The no-relata demand is grammatically motivated. When a philosopher asks 'what are the relations relations of?', the question feels mandatory. It feels like a logical requirement. But it is also a grammatical one. The question is formulated in subject-predicate structure: 'relations' is the subject, 'of' indicates the predicate-relation, and the question demands an object to complete the sentence. A grammar that treated relations as primitive would not generate this question in the same form. The no-relata demand, addressed in Chapter 2, is therefore not only a philosophical objection but a grammatical reflex – the encoding protocol insisting that its own structural assumptions be satisfied.

This does not refute the objection. Arguments refute objections; grammatical observations explain their persistence. But understanding why the demand feels mandatory – why it seems like an inescapable requirement of reason rather than a feature of a particular encoding system – is important for assessing its force. The demand has the weight of grammatical inevitability behind it. Recognising this does not dissolve the demand, but it does demote it from a logical necessity to a structural preference.

The mind-world problem is partly a grammar problem. As noted in Section 7.3, the sentence 'the mind perceives the world' installs the very bifurcation MEOW rejects. Every attempt to articulate MEOW's alternative must work against this grammatical current. 'Encounter-events are ontologically primary' still uses a noun-subject ('encounter-events') and a predicate ('are primary'), which risks reinstalling the substance architecture at the level of expression even as the content of the sentence denies it. This is not a refutation of the framework. It is a structural difficulty of articulation – and recognising it as such prevents the difficulty from being mistaken for an incoherence.

Philosophical writing is ontologically non-neutral. Every sentence in this monograph carries ontological commitments that its grammar imposes independently of its content. The framework can be articulated in Indo-European languages – it is being articulated in one now – but the articulation requires constant vigilance against the encoding biases of the medium. The reader is asked to notice, where possible, the gap between what the sentences say and what the grammar presupposes. That gap is itself evidence for the ontological grammar thesis.

7.7 The Bandwidth Problem: Where Synchronisation Fails

The synchronisation protocol has limits. These limits are not incidental – not failures of vocabulary, education, or effort that could in principle be overcome. They are structural: consequences of the protocol's operating conditions that cannot be eliminated without changing the nature of the protocol itself.

Language succeeds as a synchronisation mechanism under specific conditions:

- The constraint-patterns being communicated are sufficiently stable – slow-changing enough to be captured by a discrete encoding.
- The encountering systems share enough mediational overlap – similar enough perceptual, cognitive, and cultural organisation that the compressed signal can be approximately decompressed at the receiving end.
- The domain admits of repeatable stabilisation – the encounter-patterns can be categorised, named, and re-encountered with enough consistency to sustain linguistic reference.

When these conditions are met, language works remarkably well. Scientific terminology, technical vocabularies, legal language, and everyday object-reference all succeed because the constraint-patterns they encode are stable, the mediational overlap between users is high, and the domains permit consistent restabilisation.

When these conditions are not met, language fails – not catastrophically, but systematically. The failures cluster in predictable domains:

High fluidity. Where constraint-patterns change faster than linguistic categories can track, synchronisation degrades. Emotional states, aesthetic experiences, phenomenal qualities, and rapidly shifting social dynamics all

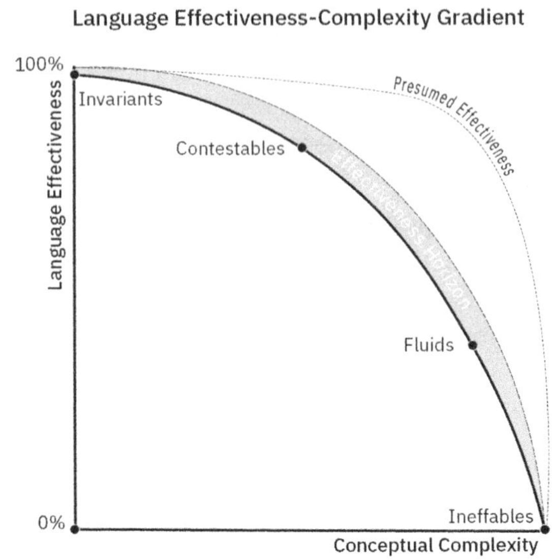

Figure 7.1: Language Effectiveness-Complexity Gradient

resist stable linguistic encoding because the patterns are in motion. By the time the encoding is complete, the pattern has changed. Language captures a snapshot; the phenomenon is a film.

Weak convergence. Where the mediational architectures of sender and receiver diverge significantly, the decompression at the receiving end produces a stabilisation that poorly approximates the original. Cross-cultural communication about values, experiences of suffering or transcendence, and discourse between radically different disciplinary frameworks all exhibit weak convergence. The words are transmitted successfully. The stabilisations they reconstruct are not the same.

Radical mediational asymmetry. Where one encountering system operates under mediational conditions fundamentally unlike the other's, the synchronisation protocol approaches its limits. Attempts to communicate the experience of synaesthesia, the phenomenology of psychosis, or – pertinently – the 'inner life' of a non-biological cognitive system all face this problem. The encoding protocol presupposes a degree of shared mediational structure that is absent.

Low discretisation tolerance. Where the constraint-patterns resist decomposition into the discrete, bounded categories that linguistic grammar requires, encoding introduces distortion. Continuous gradients, entangled phenomena, and holistic configurations all resist the discretising pressure of noun-based grammar. The encoding forces a carving that the phenomenon does not naturally support.

These failure modes are not bugs. They are structural consequences of the protocol's design. A finite-bandwidth synchronisation system that operates through discrete compression within a noun-dominant grammatical architecture will necessarily fail in domains where the phenomena are fluid, the mediational overlap is low, and the constraint-patterns resist discretisation.

7.8 The Language Insufficiency Hypothesis

The Language Insufficiency Hypothesis, developed in detail elsewhere,[7] can now be restated as a consequence of MEOW's architecture rather than an independent thesis.

If encounter is always mediated, and language is a finite-bandwidth protocol operating within that mediation, then there will necessarily be domains where the protocol's resolution is insufficient to synchronise stabilisations across agents. This insufficiency is not a contingent feature of particular languages or particular speakers. It is a structural feature of any discrete encoding system operating under the conditions described in Section 7.7.

The LIH can be stated in compressed form:

For any finite-bandwidth linguistic protocol L operating across mediational architectures M_1 and M_2, there exist domains D in which the constraint-patterns of encounter resist stable compression into L with sufficient fidelity to achieve reliable synchronisation between M_1 and M_2.

Or, less formally: there are things that language cannot say – not because speakers are inarticulate, but because the encoding protocol's resolution is insufficient for the domain. The domains in question are precisely those

7 Willis, B. (2026). *A Language Insufficiency Hypothesis*. Philosophics Press; see Bibliography.

characterised by the failure modes identified in Section 7.7: high fluidity, weak convergence, radical mediational asymmetry, and low discretisation tolerance.

Figure 7.1 does not itself depict the ontological mechanisms that dominate the *Contestables* and *Fluids* regions. It tracks the protocol's observable performance: how well stabilisations synchronise as complexity rises. The further claim of this chapter is that, in the middle regions, effectiveness degrades not only because of finite bandwidth but because the protocol's ontological grammar actively mis-shapes what it encodes. The result is not merely vagueness but systematic false convergence: the appearance of shared meaning where underlying stabilisations are incommensurate.

The relationship between conceptual complexity and linguistic effectiveness is represented in Figure 7.1, adapted from the original LIH paper. The diagram borrows its structural logic from the economist's Production Possibility Frontier, though without the trade-off that characterises the PPF: movement along the complexity axis does not purchase gains elsewhere but simply registers declining protocol performance. Four category markers locate regions along the gradient. Invariants – stable, highly discretisable constraint-patterns such as logical constants and physical measurements – occupy the high-effectiveness region where synchronisation succeeds reliably. *Contestables* – moral concepts, political categories, aesthetic judgements – sit where convergence weakens and mediational overlap between agents becomes insufficient for stable synchronisation. *Fluids* – emotional states, phenomenal qualities, rapidly shifting social dynamics – occupy the region where the protocol's bandwidth is manifestly inadequate. *Ineffables* – experiences and constraint-patterns that resist linguistic encoding entirely – mark the structural limit where effectiveness approaches zero. The shaded region between the *Effectiveness Horizon* (the actual curve) and the *Presumed Effectiveness* line (the dashed curve) represents the overestimation gap: the domain in which language is assumed to function but structurally cannot. It is in this gap that philosophical impasse, institutional dysfunction, and the reification of heuristic categories most characteristically occur. Figure 7.1 does not itself depict the ontological mechanisms that dominate the *Contestables* and *Fluids* regions. It tracks the protocol's observable performance: how well stabilisations synchronise as complexity rises. The further claim of this chapter is that, in the middle regions, effectiveness degrades not only because of finite bandwidth but because the protocol's ontological grammar actively mis-shapes what it encodes. The result is not merely vagueness but systematic false convergence: the appearance of shared meaning where underlying stabilisations are incommensurate.

The consequences ramify across the monograph's concerns.

For philosophy: Many philosophical impasses may reflect not genuine metaphysical depth but the protocol's bandwidth limitations. Debates that seem interminable – the nature of consciousness, the meaning of 'good', the essence of art – may persist not because the questions are unanswerable but because the linguistic medium through which the answers must be articulated lacks the resolution to stabilise convergent responses. The Language Insufficiency Hypothesis does not resolve these debates. It reframes their persistence as partly a feature of the medium rather than entirely a feature of the subject matter.

For governance and institutional discourse: Institutions operate through language. Law, policy, regulation, and administrative procedure are all linguistic artefacts – stabilisations encoded in the protocol and enforced through second-order constraint systems (developed in Chapter 8). When the phenomena being governed resist stable linguistic encoding – when values shift, when social dynamics outpace categorial frameworks, when the governed population's mediational architectures diverge significantly – the institutional apparatus loses grip. The language continues to function syntactically. It ceases to synchronise effectively. The result is institutional dysfunction that is experienced as political failure but is, at root, a bandwidth problem.

For the ontological grammar thesis: The LIH and the ontological grammar thesis are complementary rather than competing claims. Ontological grammar explains what language imposes – the structural commitments that shape every synchronisation act. The LIH explains where language fails – the domains in which the protocol's bandwidth is insufficient regardless of its grammatical structure. Together, they provide a comprehensive account of language's role in the architecture of encounter: it synchronises, it shapes, and it has limits. None of these features is incidental. All three are consequences of the protocol's operating conditions.

7.9 Ontological Grammar and the Shape of the Gradient

The Language Effectiveness-Complexity Gradient (Figure 7.1) represents the structural limits of the synchronisation protocol. But the shape of the curve – its rate of degradation, its specific profile across the *Contestables* and *Fluids* regions – is not determined by bandwidth alone. A second factor operates in these middle regions, and it is the ontological grammar of the encoding protocol itself.

The bandwidth limitation is straightforward: as conceptual complexity increases, the protocol's finite resolution becomes insufficient to capture the constraint-patterns with adequate fidelity. This is a generic limitation that would apply to any finite encoding system regardless of its grammatical structure.

But the ontological grammar introduces a further source of degradation. The encoding protocol is not ontologically neutral – as Sections 7.3 through 7.6 have argued, Indo-European grammar imposes substance-ontological commitments at the level of syntax. When the phenomena being encoded are themselves substance-shaped – discrete, bounded, persistent, property-bearing – this grammatical architecture introduces minimal distortion. The ontological commitments of the grammar align with the structure of the domain. Invariants sit in this region: physical constants, logical relations, well-individuated objects. The protocol performs near its theoretical maximum because the encoding structure and the phenomenon share the same ontological shape.

When the phenomena are not substance-shaped – when they are processual, relational, fluid, continuously graded, or resistant to discretisation – the grammar's ontological commitments become a source of active distortion. The protocol forces durative phenomena into noun-shaped containers. It imposes subject-predicate structure on patterns that lack that architecture. It discretises gradients that are continuous. The result is not merely loss of resolution but misrepresentation: the encoding produces a stabilisation that is ontologically misaligned with the constraint-pattern it purports to transmit. *Contestables* and *Fluids* sit in this region. Moral concepts, aesthetic judgements, phenomenal qualities, and rapidly shifting social dynamics are all poorly served not just because they are complex but because they are the wrong shape for the encoding protocol's grammatical commitments.

Two sources of degradation therefore operate simultaneously in the middle and lower regions of the gradient. The first is bandwidth: the protocol's finite resolution cannot capture the complexity. The second is ontological mismatch: the grammar's substance-commitments impose the wrong categorial structure, and the distortion compounds the bandwidth loss. A synchronisation protocol with different ontological grammar – a durative or relational encoding architecture – would not eliminate the bandwidth limitation, but it would push the effectiveness curve outward in the *Contestables* and Fluids regions by reducing the distortion introduced by the encoding structure, though it would not abolish the horizon and may introduce different distortions of its own.

A further clarification is needed. The gradient is population-relative: it reflects the average synchronisation performance across agents whose ontological grammars are not identical. A sufficiently homogeneous group – one sharing metaphysical reflexes, moral background assumptions, and interpretive defaults – would exhibit a higher and flatter curve in the *Contestables* region, because the decompression targets are structurally similar. But homogeneity reduces ontological mismatch distortion; it does not eliminate structural bandwidth limits. The sharpest declines in the *Contestables* region correlate not merely with conceptual complexity but with ontological divergence among agents: where decompression targets differ fundamentally, apparent linguistic convergence masks deep structural misalignment. This is a descriptive observation about encoding protocols under population variance. It is not a normative claim about the desirability of homogeneity.

This is why the gradient cannot reach 100% effectiveness even in principle for these domains. The bandwidth ceiling is one constraint. The ontological grammar is another. And the two interact: the grammar determines how the available bandwidth is allocated, and if the allocation is structurally misaligned with the phenomena, the effective bandwidth is lower than the nominal bandwidth. The protocol loses capacity to the encoding overhead of forcing the wrong ontological shape onto the data.

The integration of these two sources of degradation unifies the LIH and the ontological grammar thesis into a single account of the curve's shape. The LIH identifies the structural limits of the protocol. The ontological grammar thesis identifies the structural biases of the protocol. The unified claim is that the biases contribute to the limits: the curve degrades faster than it would under a less distortive encoding architecture because the grammar is consuming bandwidth on ontological overhead.

An illustrative case makes the mechanism concrete. Consider two communities attempting to synchronise on the concept of 'retributive justice'. The term transmits successfully at the syntactic level – both parties hear the words, parse the sentence, recognise the vocabulary. The synchronisation protocol appears to function. Yet the stabilisations reconstructed at each end are fundamentally different, because the ontological commitments underwriting the term diverge. One community's default ontological framing treats retributive justice against a background of individual moral agency as primitive: persons are substances bearing moral properties, and punishment is a response to a property – guilt – inhering in a substance – the offender. The other community's default framing treats it against a background of relational harm: offending is a disruption of constraint within a social field, and justice is the restabilisation of that field. Both communities use the same phrase. Neither means the same thing by it. The disagreement appears to be about justice. It is actually about the ontological grammar in which justice is encoded – about underlying commitments to concepts such as fairness, desert, and the inherent moral status of persons that are grammatically presupposed rather than explicitly argued.

This is not ambiguity. Ambiguity is a surface phenomenon that can be resolved by definition or stipulation. What is occurring is structural incommensurability at the ontological level: the encoding protocols are incompatible not because the vocabulary differs but because the grammars impose different ontological architectures on the same domain. The concept sits squarely in the *Contestables* region of the effectiveness gradient, and it illustrates why that region is so treacherous. Language does not merely lose resolution there. It appears to be working – the words transmit, the sentences parse, the conversation continues – whilst the underlying synchronisation has silently failed. The protocol produces false positives: apparent communication where actual convergence is absent. And because both parties assume the protocol is functioning – the Presumed Effectiveness curve – the failure is attributed to the interlocutor's stubbornness, bad faith, or irrationality rather than to a structural feature of the encoding system.

This is the overestimation gap in action. And it is where political and institutional discourse most characteristically breaks down – not in the *Ineffables* region, where everyone acknowledges language's limits, but in the *Contestables* region, where the limits are invisible precisely because the protocol appears to be working.

7.10 The Fluency Tax

A final observation draws together the threads of this chapter and connects them to the monograph's broader methodological commitment.

Substance ontology is fluent. It flows with the grammar. Subject-predicate sentences carry it without friction. 'The mind perceives.' 'Objects exist.' 'Causation produces effects.' Each sentence is clear, compact, and grammatically natural. The encoding protocol rewards substance-ontological framings with ease of articulation.

Durative, interaction-first ontology pays a fluency tax. 'Temporarily stabilised constraint-clusters within scale-dependent encounter-regimes exhibit convergent invariance across mediational variation' says something precise and defensible. It also makes the reader work. The encoding protocol penalises it with cumbersomeness.

This asymmetry is not a philosophical argument for substance ontology. It is a feature of the encoding medium. The grammar makes one picture easy and the other hard. Cognitive economy – the principle of least effort – does the rest. Speakers default to the picture that flows. The picture that flows happens to be substance ontology. Substance ontology therefore feels like common sense, not because it is correct but because the grammar rewards it with fluency.

The present monograph accepts this fluency tax deliberately. It chooses structural fidelity over cognitive economy.[8] The framework is more cumber-

---

8 The concept of a 'fluency tax' has a structural analogue in the economics of communication. Zipf's (1949) principle of least effort predicts that speakers will favour shorter, more frequent, and more easily processed expressions – which, within the present framework, means expressions whose ontological grammar aligns with the default categorial structure of the language. Expressions that work against the grain of the grammar's ontological commitments are, in Zipfian terms, more costly to produce and process. The fluency tax is therefore not a metaphor but a predictable consequence of the economics of articulation: substance-ontological framings are Zipf-optimal in Indo-European languages, and relational-ontological framings are not. The monograph's acceptance of this cost is

some than substance ontology, and the cumbersomeness is the cost of refusing a lossy compression. Where substance thinking reifies stabilisations into self-subsisting entities and generates pseudo-problems – the no-relata demand, the mind-world bifurcation, the hard problem of consciousness – the durative alternative preserves the structure of the phenomena at the cost of articulatory ease.

This is not an apology. It is a methodological declaration. The cost of structural fidelity is articulatory resistance. The alternative is ease purchased through distortion. The reader is invited to notice the tax being paid, and to consider whether the fidelity it purchases is worth the price.

7.11 What This Chapter Has Established

Language is a finite-bandwidth synchronisation protocol that aligns predictive stabilisations across distinct mediational architectures. It succeeds where constraint-patterns are stable, mediational overlap is sufficient, and the domain admits of discrete restabilisation. It fails – systematically and structurally – where these conditions are not met.

The grammatical structure of natural language is not ontologically neutral. Indo-European subject-predicate grammar scaffolds substance ontology by making object-first thinking fluent and relational or durative thinking cumbersome. This ontological grammar explains the persistence of substance thinking even after its philosophical inadequacies have been diagnosed. The explanation is not deterministic: speakers can and do think against their grammars. But the grammar taxes alternatives, and cognitive economy favours the path of least resistance.

The Language Insufficiency Hypothesis follows from MEOW's architecture as a structural consequence: any finite-bandwidth protocol operating across non-identical mediational systems will necessarily fail in domains of high fluidity, weak convergence, or radical mediational asymmetry. These failures are not defects of language but features of its operating conditions. Moreover, the rate of degradation across the effectiveness gradient is shaped not only by bandwidth limitations but by the ontological mismatch between the protocol's grammatical commitments and the structure of the phenomena being encoded. The biases of the protocol contribute to its

a commitment to structural fidelity over communicative efficiency – a trade-off that the effectiveness gradient itself would predict will limit the monograph's synchronisation range.

limits: where grammar and phenomenon are misaligned, the effective bandwidth is lower than the nominal bandwidth.

The chapter has also identified the fluency tax that durative ontology pays within noun-dominant grammar, and has declared the monograph's acceptance of that tax as a deliberate methodological commitment.

These observations have immediate consequences for social ontology – for understanding how institutions, norms, and shared categories are stabilised through linguistic coordination and destabilised when the protocol's bandwidth is exceeded. That is the subject of Chapter 8.

EXTENSIONS

III

8

Social Ontology

Second-Order Constraint Systems

The judges of normality are present everywhere.
Les juges de normalité sont partout.

— Michel Foucault, *Discipline and Punish*

It is often safer to be in chains than to be free.
Es ist oft sicherer, in Ketten zu sein, als frei zu sein

— Franz Kafka, *The Trial*

8.1 Introduction: The Institutional Question

The previous chapter established that language is a finite-bandwidth synchronisation protocol whose grammatical structure imposes ontological commitments on everything it encodes. The natural question – the one the reader has been waiting for – is: what happens when entire societies attempt to coordinate through this protocol?

The answer is institutions.

Institutions – legal systems, governments, markets, religions, bureaucracies, educational structures – are the large-scale coordination mechanisms through which human societies organise collective action. They are also, without exception, built from language. Laws are sentences. Policies are documents. Constitutions are texts. Contracts are agreements encoded in the synchronisation protocol. Markets operate through linguistically mediated price signals, regulations, and contractual structures. Even institutions that appear to be primarily material – armies, hospitals, factories – are organised through linguistic coordination: orders, procedures, diagnoses, schedules.

If language is a finite-bandwidth synchronisation protocol with structural limits and ontological biases, then institutions built from language inherit those limits and biases. This is not a metaphorical claim. It is a structural one. Institutional coordination is linguistic coordination at scale, and the failure modes identified in Chapter 7 – bandwidth degradation, ontological mismatch, false convergence – operate within institutional settings with the same structural predictability as they do in interpersonal communication. They simply operate at larger scales and higher stakes.

This chapter develops a MEOW-derived account of social ontology. It argues that social constructs – money, law, property, institutional roles, political categories – are not illusions and not mere conventions. They are second-order constraint systems: stabilised patterns of coordination that alter real interactional trajectories by changing what organisms can do, must do, and are prevented from doing. They are real in the sense that matters: they constrain encounter. But they are also linguistically mediated, which means they inherit the structural limitations of the protocol through which they are constituted and maintained.

8.2 Searle and the Limits of Declarative Ontology

The most influential contemporary account of institutional reality is John Searle's theory of institutional facts (Searle, 1995, 2010). Searle's central claim is elegant: institutional reality is created by declarations – speech acts that bring institutional facts into existence by collectively recognising them as such. Money is money because we collectively treat it as money. A border is a border because we collectively recognise it as one. The formula is: 'X counts as Y in context C'. A piece of paper (X) counts as legal tender (Y) in the United States (C).

Searle's account has two significant virtues. First, it takes social reality seriously as ontologically real without reducing it to physical reality. Money is not 'just paper'. It is paper that has been assigned a status function by collective recognition. Second, it locates the mechanism of institutional creation in language – specifically, in the capacity of language to create facts through declaration. For Searle, language is not merely a tool for describing institutional reality. It is the mechanism that constitutes it.

MEOW agrees with the second point and departs from the first. The disagreement concerns not whether language constitutes institutional reality but what the constitutive mechanism ultimately depends upon.

The departure concerns the mechanism of constitution. Searle treats collective recognition as the ontological ground of institutional reality. On his account, institutions exist because and only because people collectively accept or recognise them as existing. This makes institutional reality ontologically dependent on collective intentionality – on the shared mental states of the participants.

MEOW reframes this. Institutions are not sustained by collective mental states. They are sustained by coordinated behaviour, enforcement mechanisms, and material embedding[1] – that is, by second-order constraint

1 The critique of Searle's reliance on collective intentionality is not unique to MEOW. Latour (2005) argues that social facts are sustained by networks of human and non-human actors – material artefacts, documents, technologies – rather than by shared mental states. Epstein (2015) has argued that Searle's account is insufficiently attentive to the material and structural conditions that ground institutional facts, proposing a 'grounding' framework that locates institutional reality in social practices and their enabling conditions rather than in collective recognition. MEOW's departure from Searle is adjacent to both critiques but grounded in the specific architecture of encounter: the mechanism is constraint (coordinated behaviour that alters interactional trajectories) rather than either Latour's networks or Epstein's social grounding. The convergence of independent critiques on Searle's

systems that alter real interactional trajectories regardless of whether any particular participant 'recognises' them. Money constrains your behaviour whether or not you have philosophical views about its ontological status. A border constrains movement whether or not the person crossing it accepts the legitimacy of the state that drew it. The constraint is real. It operates through encounter. It does not require collective intentionality as its ontological ground.

This is not a minor adjustment. It is a structural relocation. Searle grounds institutional reality in minds (collective recognition). MEOW grounds it in constraint (coordinated behaviour that alters interactional trajectories). Searle's account inherits the substance ontology that MEOW rejects: it presupposes independently existing minds that subsequently generate institutional facts through shared intentional states. MEOW derives institutional reality from the same constraint architecture that grounds physical and perceptual reality – not because institutions are physical, but because the mechanism is the same. Patterns of interaction stabilise. When they stabilise at the social scale, through linguistic coordination and material enforcement, we call them institutions.

The practical consequence is significant. On Searle's account, institutional failure is ultimately a failure of collective recognition – people stop believing in the institution, and it collapses. On MEOW's account, institutional failure is more commonly a failure of synchronisation – the linguistic protocol through which the institution coordinates can no longer carry the ontological load placed upon it. These are different diagnoses, and they point toward different explanations of why institutions persist, strain, and break.

8.3 Second-Order Constraint Systems

The concept of a second-order constraint system requires precise articulation, because it is doing specific work that distinguishes it from both Searlean institutional facts and ordinary social conventions.

A first-order constraint is a pattern of stability within encounter-events at the physical, perceptual, or cognitive level. Gravity constrains falling bodies. Spectral regularities constrain colour encounter. Cognitive architecture constrains inference. These constraints operate regardless of linguistic mediation – they are features of encounter-organisation at the biological and physical scale.

intentionalism suggests that the problem is structural rather than idiosyncratic.

As mentioned, a second-order constraint is a pattern of stability that is constituted and maintained through linguistic coordination and sustained by coordinated behaviour, enforcement mechanisms, and material embedding.[2] Property rights, legal obligations, monetary value, institutional roles, and political categories are all second-order constraints. They alter real interactional trajectories – in ways that are observable regardless of participants' metaphysical beliefs – but they do so through a layer of linguistic mediation that first-order constraints do not require.

The distinction matters because second-order constraints inherit the structural features of the protocol through which they are constituted. First-order constraints are disclosed through mediation but do not depend on language for their stability. Gravitational constraint persists whether or not anyone names it. Second-order constraints do depend on language for their stability – they must be articulated, transmitted, enforced, and periodically restabilised through the synchronisation protocol. This dependence makes them simultaneously powerful and fragile.

Powerful, because language enables coordination at scales that no other mechanism can achieve. A legal system synchronises the behaviour of millions of agents through a shared vocabulary of rights, obligations, and procedures. A monetary system coordinates economic activity across continents through linguistically mediated signals. No non-linguistic coordination mechanism has this reach.

Fragile, because the protocol on which they depend has structural limits. When the constraint-patterns that institutions attempt to stabilise sit in the Contestables or Fluids regions of the effectiveness gradient – when they involve moral judgement, value conflict, identity, fairness, or rapidly shifting social dynamics – the synchronisation protocol begins to degrade. The insti-

2 The concept of second-order constraint bears a structural resemblance to several existing frameworks in social theory, though it is not derived from any of them. Scott's (1998) Seeing Like a State analyses how institutional legibility projects – the imposition of standardised categories on complex social phenomena – both enable governance and produce systematic distortions. The parallel with MEOW's account of institutional ontological grammar is close: Scott's 'legibility' is the institutional demand for noun-shaped categories that the synchronisation protocol can process, and his catalogue of its failures corresponds to the bandwidth degradation and false convergence identified in this chapter. The difference is that Scott treats legibility as a political strategy, whereas MEOW treats it as a structural consequence of the encoding protocol's operating conditions. The failures Scott diagnoses as political are, within this framework, also – and more fundamentally – linguistic.

tution continues to operate syntactically. The words continue to circulate. But the underlying synchronisation – the alignment of stabilisations across agents – silently fails.

8.4 How Institutions Inherit Ontological Grammar

Chapter 7 established that the grammatical structure of natural language imposes ontological commitments on everything it encodes. Institutions, as linguistically constituted coordination systems, inherit these commitments.

The inheritance operates at several levels.

Legal and administrative categories are noun-shaped. The law operates through categories: person, property, intent, negligence, harm, discrimination, reasonable force. Each is a noun – a grammatically stabilised entity that the legal system treats as a discrete, bounded, identifiable thing. But many of these categories are better understood as constraint-clusters within encounter-organisation: fluid, context-dependent, and resistant to the sharp boundaries that noun-grammar demands. 'Intent' is not an object inside a skull. It is a retrospective attribution imposed on a sequence of encounter-events, compressed into a noun so that the legal system can process it. The noun makes the concept juridically operable. It also makes it ontologically misleading.

Policy language operates through high-variance coordination placeholders. Terms like sustainability, resilience, equity, innovation, and public value function within institutional discourse as what might be called high-variance coordination placeholders:[3] words that enable multiple agents to participate in the same policy conversation whilst compressing radically different stabilisations into the same term. The word synchronises the conver-

3 The phenomenon identified here as 'high-variance coordination placeholders' has been observed in several adjacent literatures under different names. Gallie (1956) introduced the concept of 'essentially contested concepts' to describe terms whose proper use is inherently disputable because the concept involves an evaluative component that reasonable people will weigh differently. Connolly (1993) extended the analysis to political concepts specifically. MEOW's treatment differs from Gallie's in a diagnostic respect: it does not attribute the contestedness to the evaluative character of the concept (an explanation that remains within substance ontology – the concept has a property, evaluativeness, that generates dispute). It attributes it to the structural conditions of encoding: the terms sit in a region of the effectiveness gradient where the protocol's bandwidth is insufficient for convergent synchronisation, and the resulting divergence is built into the protocol's operating conditions rather than into the concepts themselves.

sation. The underlying stabilisations remain misaligned. This pattern – apparent linguistic coordination where actual convergence is absent – recurs throughout institutional life and will be referred to in what follows as institutional false convergence. It is not a failure of the participants. It is a structural feature of the protocol operating in a domain where ontological divergence among agents is high and the constraint-patterns resist stable discretisation. The policy term survives precisely because its ambiguity temporarily permits multiple decompressions to coexist. Precision would expose the divergence and halt the coordination.

Political categories inherit the substance-ontological defaults of the grammar. Concepts like freedom, justice, rights, and security are treated within political discourse as if they name stable, identifiable things about which people can agree or disagree. But from within the MEOW framework, these terms are better understood as compressed references to complex, internally contested encounter-organisations – constraint-patterns whose structure varies across mediational architectures, cultural inheritances, and ontological grammars. When citizens debate 'justice', they typically assume they are arguing about the same object. They are often operating with incompatible ontological framings that the shared vocabulary conceals. The debate persists not because the participants are uniquely irrational but because the synchronisation protocol is operating beyond its effective bandwidth, in the overestimation gap between presumed and actual effectiveness.

In each case, the pattern is the same. The institution takes a fluid, relational, contextually variable phenomenon, compresses it into a noun-shaped legal or political category, and treats the resulting stabilisation as if it named a determinate thing. The grammatical form makes the category administratively usable. It also imports substance-ontological commitments that may be structurally misaligned with the phenomenon the category is meant to govern.

8.5 Institutional Persistence Through Synchronisation Repair

If institutions operate through a protocol with structural limits, how do they persist at all?

The answer is not that the protocol works better than Chapter 7 suggests.

It is that institutions have developed mechanisms for repairing synchronisation failures without acknowledging them as such. The principle can be stated directly: institutions persist not because their categories remain stable but because they continuously repair synchronisation failures produced by the protocol that constitutes them.

The most visible repair mechanism is precedent. Legal systems survive not because their categories are clear but because they continuously renegotiate the stabilisation each time a case appears. A judge confronting a novel application of 'reasonable force' is performing synchronisation repair: taking a noun-shaped category that has drifted out of alignment with the constraint-patterns it governs and locally restabilising it for the present case. Precedent is the institutional memory of these repairs – a record of how the synchronisation has been patched over time. Common law, in this light, is not a system of fixed rules but a continuous synchronisation-repair process operating under the fiction that the categories it repairs were stable all along.[4]

A second repair mechanism is institutional hermeneutics – the interpretive practices through which authorised agents (judges, regulators, administrators, committees) determine what institutional language means in contested cases. These practices are often presented as the discovery of a meaning that was already latent in the text. From the MEOW perspective, they are better understood as re-synchronisation events: moments at which the institution's linguistic apparatus is realigned with current constraint-patterns through an authoritative act of reinterpretation.

A third repair mechanism is strategic ambiguity. Institutional language is often deliberately vague – not as a defect but as a design feature. Constitutions, international treaties, and framework legislation frequently employ terms whose ambiguity is functional: it permits coordination to

4 The characterisation of legal reasoning as continuous reinterpretation disguised as consistent application has parallels in jurisprudence. Hart's (1961) distinction between the 'core' and 'penumbra' of legal concepts acknowledges that legal terms have a zone of settled application surrounded by a zone of indeterminacy. Dworkin's (1986) 'law as integrity' treats legal reasoning as constructive interpretation that aims for the best moral reading of the legal materials. MEOW's account is more structural than either: it identifies the indeterminacy not as a feature of particular legal concepts but as a predictable consequence of the encoding protocol's operating conditions. Legal concepts are indeterminate in the 'penumbra' because the constraint-patterns they stabilise sit in the Contestables region of the effectiveness gradient. The repair process (precedent, reinterpretation) is the legal system's structural response to this indeterminacy, not evidence that something has gone wrong.

continue in the absence of genuine convergence. The ambiguity absorbs ontological divergence by allowing multiple agents to decompress the same term differently whilst maintaining the appearance of agreement. This works until the divergence exceeds the ambiguity's absorptive capacity, at which point the institution enters crisis – not because something new went wrong, but because the synchronisation failure that was always present becomes visible.

These repair mechanisms are not symptoms of institutional weakness. They are the immune system of linguistically constituted coordination. They allow institutions to function in domains where the protocol's bandwidth is insufficient for stable synchronisation, by continuously patching the gaps between what the language says and what the constraint-patterns require.

But repair has limits. When the rate of change in the underlying constraint-patterns exceeds the institution's capacity for synchronisation repair – when social dynamics shift faster than precedent can track, when the ambiguity absorbs more divergence than it can structurally contain, when the authorised interpreters lose legitimacy – the institution's coordination degrades. What is experienced as political crisis, institutional failure, or democratic dysfunction is, at root, the synchronisation protocol exceeding its operating parameters.

8.6 Diagnostic: Legal Categories as Unstable Constraint-Clusters

The framework's explanatory purchase can be demonstrated through a closer examination of how legal categories function under the account developed here.

Consider the concept of negligence. In legal discourse, negligence is treated as a property – something a person either exhibited or did not. The grammatical form is substance-ontological: the defendant was negligent. The noun presupposes an identifiable, bounded state that inhered in the defendant's conduct.

But negligence, viewed through MEOW, is not a property inhering in a substance. It is a retrospective attribution applied to a sequence of encounter-events, assessed against a standard of care that is itself a second-order constraint – a linguistically stabilised norm maintained through precedent, professional practice, and institutional enforcement. The assessment involves:

- reconstructing the encounter-sequence (what happened)
- comparing it against a constraint-pattern (the standard of care)
- determining whether the deviation was sufficient to cross a threshold (the legal finding)

Each step involves synchronisation: the court must align its reconstruction of the encounter-sequence with the accounts of witnesses, experts, and documentary evidence – each of which is itself a compressed stabilisation transmitted through the protocol. The standard of care must be synchronised across judge, jury, and legal community. The threshold must be stabilised well enough to permit a binary finding (negligent / not negligent) from what is, in reality, a gradient.

The legal system manages this by grammatically stabilising negligence as a noun – a thing that can be found present or absent. The grammatical form enables the institution to function. But the functioning depends on continuous synchronisation repair: each case locally renegotiates what 'negligence' means in the context at hand, building a trail of precedent that records the sequence of repairs without admitting that repairs were needed.

The framework does not claim that the legal system is 'broken'. It claims that the legal system's characteristic mode of operation – continuous reinterpretation disguised as consistent application – is the predicted behaviour of a second-order constraint system operating through a finite-bandwidth protocol in a domain of moderate ontological complexity. The institution works. But it works by means that its own self-description does not acknowledge.

8.7 Diagnostic: Policy Language and False Convergence

Policy discourse offers a second diagnostic site, because it exhibits the false convergence phenomenon identified in Chapter 7 with particular clarity.

Consider the term resilience as it functions in contemporary governance. The word circulates freely across policy documents, strategic plans, funding applications, and political speeches. It appears to name a determinate property: the capacity of a system to absorb disruption and recover. Its grammatical form is reassuringly noun-like. Everyone agrees that resilience is desirable.

But the stabilisations compressed into the term diverge radically across agents. For an ecologist, resilience references a specific dynamic – the capacity of an ecosystem to return to a prior equilibrium after perturbation. For a financial regulator, it references capital adequacy and systemic risk absorption. For a community development officer, it references social cohesion and adaptive capacity under economic stress. For a military planner, it references infrastructure hardening and continuity of operations. Each agent decompresses 'resilience' within a different mediational architecture and against different background constraint-patterns. The word synchronises the policy conversation. The stabilisations it reconstructs are not the same.

This is not ambiguity in the ordinary sense. Ambiguity implies a determinate meaning that has been obscured. What is occurring is structural: the term functions as a coordination placeholder that permits multiple agents to participate in the same institutional process whilst compressing incompatible stabilisations into a shared vocabulary. The divergence is not a bug. It is what allows the policy to be written at all. Precision would expose the incompatibility and halt the process.

The framework predicts this behaviour. Policy language operates squarely in the Contestables region of the effectiveness gradient. The constraint-patterns it governs are normatively loaded, identity-bound, and resistant to stable discretisation. The ontological grammars of the participating agents diverge. The protocol produces false convergence – apparent agreement where actual synchronisation is absent – because the operating conditions exceed its structural parameters.

The institutional consequence is familiar: policies are adopted with broad nominal support, then fragment upon implementation as the divergent stabilisations become operationally visible. What is experienced as 'implementation failure' is, from this framework's perspective, the deferred revelation of a synchronisation failure that was present from the outset.

8.8 Diagnostic: Democratic Deliberation and the Overestimation Gap

Democratic deliberation provides a third diagnostic, and it is the one that most directly illustrates the overestimation gap identified in Figure 7.1.

The ideal of democratic deliberation presupposes that citizens can reason together about shared concerns, articulate competing views, and reach decisions through argument.[5] The presupposition is that the synchronisation protocol – public language – is adequate to the task. Citizens may disagree, but they are assumed to be disagreeing about the same things, using shared concepts, within a common framework of meaning.

The MEOW framework challenges each of these assumptions.

Citizens do not necessarily disagree about the same things. When two voters argue about 'freedom', one may be operating within an ontological framing that treats freedom as a property of individuals – a substance-bearing-attribute – whilst the other operates within a framing that treats freedom as a relational condition – a constraint-configuration within a social field. The word is identical. The stabilisations are incommensurate. The debate appears to be about the correct account of freedom. It is actually about the ontological grammar in which freedom is encoded.

The concepts are not shared in the way the deliberative ideal requires. Terms like justice, equality, security, and rights function within political discourse as Invariants-region vocabulary – stable, well-defined, suitable for rational argument. But they sit in the Contestables region of the gradient, where the protocol's bandwidth is insufficient for reliable synchronisation across agents with divergent ontological grammars and mediational histories. The presumed effectiveness curve extends into this territory. The actual effectiveness curve does not.

5 The deliberative ideal has its most rigorous philosophical formulation in Habermas's (1984, 1996) theory of communicative action and discourse ethics. Habermas argues that rational consensus is achievable under conditions of undistorted communication – the 'ideal speech situation' in which all participants have equal voice, argument proceeds by the force of the better reason, and external coercion is absent. MEOW's framework suggests that the ideal speech situation, even if politically achievable, would not resolve the synchronisation problem. Undistorted communication addresses power asymmetries but not bandwidth limitations. Participants freed from coercion still operate through a finite-bandwidth protocol whose ontological grammar imposes structural biases and whose resolution degrades in the Contestables region. The deliberative ideal is not wrong. It is incomplete: it diagnoses one source of communicative failure (power) and overlooks another (protocol structure).

The overestimation gap is where democratic frustration lives. Citizens assume the protocol is working – the words transmit, the sentences parse, the arguments are grammatically well-formed – and therefore attribute communicative failure to the interlocutor: bad faith, stupidity, wilful ignorance, moral deficiency. The framework suggests a less dramatic but more structural explanation. The protocol is operating beyond its parameters. The failure is in the encoding, not in the interlocutors.

This is not a counsel of despair about democracy. It is a diagnostic. Democratic deliberation works tolerably well in domains where the operating conditions favour synchronisation – where constraint-patterns are stable, mediational overlap is high, and the issues admit of discrete stabilisation. It degrades predictably in domains where these conditions are absent. Understanding why it degrades – and understanding that the degradation is structural rather than moral – is a precondition for any serious attempt to improve it.

8.9 The Fragility of Second-Order Constraint Systems

The diagnostics of Sections 8.6 through 8.8 converge on a common structural observation. Institutions built from language are simultaneously powerful and fragile – powerful because language enables coordination at extraordinary scales, fragile because the protocol on which they depend has structural limits that the institutions themselves cannot acknowledge without undermining their own authority.

The fragility has a specific character. It is not the fragility of physical systems (bridges that collapse under excess load). It is the fragility of synchronisation systems: the gap between what the protocol appears to be doing (coordinating shared understanding) and what it is actually doing (circulating compressed stabilisations that are decompressed differently by different agents) can widen gradually without becoming visible, until a threshold is crossed and the divergence becomes operationally undeniable.

This fragility is amplified by three interacting mechanisms identified across the preceding chapters:

Bandwidth degradation. As the conceptual complexity of the domain increases, the protocol's resolution decreases. Institutions governing simple, well-discretised phenomena (traffic regulations, measurement standards, basic property rules) are robust. Institutions governing complex, normatively loaded phenomena (constitutional rights, social policy, international law) are structurally fragile – not because the framers were incompetent but because the protocol is operating near or beyond its limits.

Ontological grammar distortion. The grammatical structure of the protocol forces fluid, relational, processual phenomena into noun-shaped categories. Institutions inherit this distortion and treat the resulting nouns as stable objects of governance. When the phenomena shift – when social dynamics change faster than the categories can track – the institution continues to govern the nouns whilst the phenomena move elsewhere.

Population ontological variance. In pluralistic societies, the agents coordinated by an institution operate with divergent ontological grammars – different default framings of personhood, agency, harm, fairness, and value. The protocol produces false convergence across this divergence. The institution appears to coordinate. The underlying stabilisations are misaligned. The misalignment becomes visible only when the institution attempts to enforce a specific interpretation, at which point the apparent consensus fractures.

These three factors are not independent. They interact. Ontological grammar distortion increases the effective bandwidth cost of encoding. Population variance amplifies the divergence in decompression. Bandwidth degradation reduces the protocol's capacity to absorb either. The result is a system that is structurally prone to silent synchronisation failure – coordination that appears to function whilst the underlying alignment degrades.

8.10 What Institutions Cannot Say About Themselves

A final structural observation follows from the account developed in this chapter, and it concerns a limitation that is rarely acknowledged in institutional theory.

Institutions cannot, in general, describe their own synchronisation failures in the language through which they operate. The protocol that constitutes the institution is the same protocol that would need to diagnose its limitations. But the diagnosis requires articulating precisely the features of the protocol that the protocol's own structure renders invisible – its ontological grammar, its bandwidth limits, its false convergence patterns.

The institution's self-description is conducted in the same noun-dominant, substance-presupposing language that produces the distortions being diagnosed.

This is not a conspiracy. It is a structural feature of linguistically constituted systems. A legal system that acknowledged its categories are unstable constraint-clusters subject to continuous synchronisation repair would undermine the fiction of consistent application on which its authority depends. A policy framework that acknowledged its key terms function as coordination placeholders rather than determinate concepts would lose the appearance of precision that legitimates its operation. A democratic system that acknowledged its deliberative vocabulary operates beyond the protocol's effective bandwidth would call into question the ideal of rational public discourse on which its legitimacy rests.

Institutions therefore have a structural incentive to treat their linguistic categories as stable, their coordination as genuine, and their failures as aberrations rather than predicted features of the system. The MEOW framework does not blame institutions for this. It explains why the self-description is structural rather than optional. The protocol's limitations are invisible from within the protocol, in the same way that the ontological commitments of a grammar are invisible to speakers who have never encountered an alternative grammar.[6]

8.11 Toward Normativity

The account developed in this chapter has been deliberately descriptive. It has explained how institutions are constituted, how they persist, how they inherit the limitations of the linguistic protocol through which they operate, and why they fail in structurally predictable ways.

6 The structural invisibility of the protocol's limitations to the institutions that operate through it has parallels in several traditions. Wittgenstein's (1953) observation that the limits of language are the limits of the world, cited in the Chapter 7 epigraph, implies that what cannot be said within a language-game cannot be diagnosed from within it either. Foucault's (1975) analysis of disciplinary power identifies how institutional categories (normality, deviance, health, criminality) shape perception and conduct without being perceived as impositions – the mechanism operates through its invisibility. MEOW's account is more specific than Foucault's: it identifies the encoding protocol as the structural source of the invisibility and the ontological grammar as the mechanism through which institutional categories acquire their apparent naturalness. Foucault diagnoses the effect. MEOW identifies the infrastructure.

It has not addressed the question of when institutional constraint becomes obligation – when a second-order constraint system acquires normative force rather than merely behavioural regularity. The framework can describe what constraints obtain and what coordination mechanisms sustain them. Whether it can explain why some constraints are binding – why they carry moral or political authority rather than merely causal efficacy – is a question that the present architecture has not yet answered.

The ingredients of an answer are visible. Organismal persistence generates something like interests. Coordination failure generates something like the need for obligation. Second-order enforcement generates something like institutional authority. But 'something like' is not 'is', and the gap between descriptive constraint and normative bindingness has not been closed by the machinery developed so far. That gap is the subject of Chapter 11.

8.12 What This Chapter Has Established

Social constructs are not illusions and not mere conventions. They are second-order constraint systems – stabilised patterns of coordination constituted through linguistic mediation and sustained by coordinated behaviour, enforcement mechanisms, and material embedding. They are real because they constrain encounter. They are fragile because they depend on a protocol with structural limits.

Searle's institutional ontology correctly identifies language as the constitutive mechanism of institutional reality but incorrectly grounds that reality in collective intentionality. MEOW grounds it in constraint: institutions alter real interactional trajectories regardless of whether participants recognise or accept them.

Institutions inherit the ontological grammar of the linguistic protocol through which they are constituted. Legal categories, policy language, and political concepts all bear the substance-ontological commitments of noun-dominant grammar, treating fluid relational phenomena as stable objects of governance. Institutions persist through synchronisation repair mechanisms – precedent, authoritative reinterpretation, strategic ambiguity – that patch the gap between what the protocol says and what the constraint-patterns require.

Institutional failure in complex, normatively loaded domains is structurally predictable: it is the expected behaviour of a finite-bandwidth synchronisation protocol operating beyond its parameters, under ontological grammar distortion, across a population with divergent ontological framings. Institutional success and synchronisation failure are not mutually exclusive; the repair mechanisms described above allow institutions to function despite persistent underlying divergence. The failure lies primarily in the encoding, not in the participants.

The framework's diagnostic power has been demonstrated across three domains – legal categories, policy language, and democratic deliberation – each of which sits in the Contestables region of the effectiveness gradient and exhibits the characteristic pattern: apparent linguistic coordination masking silent synchronisation failure.

What the framework has not yet addressed is normativity: whether the architecture of constraint can ground obligation, not merely regularity. That is the frontier identified in Chapter 11, and it remains open.

HORIZONS

IV

9

Realism

Cross-Perspectival Convergence and the Invariant Anchor

There are real things, whose characters are entirely independent of our opinions about them.

— Charles S Peirce, 'The Fixation of Belief'

Reality is not always probable, or likely.

— Jorge Luis Borges

9.1 Introduction: The Realist Reckoning

The framework developed across the preceding chapters has been building a debt. It has rejected substance ontology. It has treated objects as scale-dependent stabilisations. It has characterised knowledge as stabilisation under encounter pressure rather than correspondence to a mind-independent world. It has shown that language imposes ontological commitments through its grammatical structure. It has located meaning in encounter-relative affordance gradients. It has demonstrated that institutions inherit the structural limitations of the protocol through which they are constituted.

A sceptical reader – patient, sympathetic, but unconvinced – is entitled to ask: is there anything left of realism?

The framework has incurred an audit. Chapter 9 is where it shows its receipts. Anticipatory passages in Chapters 2 and 3 have signposted this defence. The present chapter delivers it.

The answer is: yes. MEOW is a realist framework. But the realism it defends is not the realism of substances, not the realism of correspondence, and not the realism that requires unmediated access to a mind-independent world. It is convergence realism – realism grounded in the invariance of relational constraint across mediational variation. This chapter consolidates that position and defends it against the strongest versions of the objections it faces.

9.2 What Realism Requires

Before defending MEOW's realism, it is worth stating what any realism must minimally provide.

A realist position must hold, at minimum, that there is something about reality that is not up for negotiation. Something persists regardless of how it is described, categorised, conceptualised, or mediated. This 'something' is what distinguishes realism from constructivism (where reality is constituted by our conceptual or social practices) and from idealism (where reality is dependent on mind or experience).

The traditional way to provide this is through appeal to mind-independent objects – things that exist as they are regardless of whether anyone perceives, thinks about, or describes them. MEOW cannot take this path. It has rejected substances. It has dissolved the mind-world bifurcation. It has treated objects as encounter-relative stabilisations. The traditional realist anchor is unavailable.

But the requirement – that something is not up for negotiation – can be met without substances. What is not up for negotiation, within MEOW, is relational constraint. The patterned invariance, resistance, and stability that persist across encounters organised by different mediational architectures. Gravity constrains regardless of conceptual framework. Toxicity constrains regardless of linguistic category. The decay constant of carbon-14 constrains regardless of the cultural formation of the physicist measuring it. These constraints are not produced by encounter. They are disclosed by encounter – structured through mediation but not manufactured by it.

This is the invariant anchor. It is what makes MEOW a realism rather than a constructivism. The principle can be stated early: filtering does not imply fabrication. Mediation shapes how constraint is disclosed, but it does not manufacture what is disclosed. The evidence for this is convergence – and, crucially, the capacity for reliable intervention. When differently mediated encounter-systems not only converge on the same constraint-patterns but successfully manipulate reality on the basis of those patterns – antibiotics kill bacteria, semiconductors conduct, orbital mechanics lands spacecraft – the fabrication hypothesis becomes untenable.[1] The full defence of this principle is developed in Section 9.7. The mechanism through which invariance is justified is cross-perspectival convergence.

9.3 Cross-Perspectival Convergence

The central argument for MEOW's realism is convergence.

When distinct mediational architectures – different species, different instruments, different conceptual frameworks, different cultural formations – converge on the same constraint-patterns, that convergence is evidence of invariant structure rather than coincidence of projection.

1 The appeal to successful intervention as evidence for realism has its most rigorous formulation in Hacking's (1983) entity realism. Hacking argues that we are warranted in believing in theoretical entities when we can manipulate them to produce predicted effects: 'if you can spray them, they are real'. MEOW's convergence realism is structurally adjacent but differs in what it takes to be real. Hacking's entity realism retains a commitment to entities – to things that are sprayed, manipulated, and intervened upon. MEOW replaces entities with constraint-patterns: what successful intervention demonstrates is not the existence of a substance but the invariance of a constraint-pattern that can be reliably exploited across mediational contexts. The practical difference is negligible. The ontological difference is significant: MEOW's realism does not require commitment to the entity-ontology of the intervening theory, only to the constraint-patterns that the intervention exploits.

The argument is not that convergence proves mind-independent existence in the substance-ontological sense. MEOW does not claim access to things-in-themselves. The argument is that convergence across mediational variation is the strongest available evidence for structural invariance – for the existence of constraint-patterns that are not artefacts of any particular mediational system.

Consider a series of cases, ordered by increasing mediational diversity.

Intra-species convergence. Two human observers with normal vision, viewing the same scene from different positions, converge on the spatial layout of the environment. The convergence is not surprising – their mediational architectures are nearly identical. But it confirms that the constraint-patterns of the scene are stable across positional variation.

Cross-instrumental convergence. A thermometer and an infrared camera, operating through entirely different physical mechanisms, converge on the temperature of a surface. The convergence is more evidentially powerful because the mediational architectures diverge significantly. Neither instrument 'sees' the same way. Their agreement is explained most parsimoniously by the existence of a shared constraint-pattern (thermal energy distribution) that both instruments disclose, differently but convergently. A sceptic might object that convergence reflects shared bias – common coupling to the same physical environment rather than evidence of invariant structure. The decisive reply is that convergence is not mere agreement. It is agreement under heterogeneity plus successful intervention: the constraint-pattern, once stabilised, enables reliable manipulation (prediction, control, engineering) across contexts. An air conditioning system designed on the basis of these convergent measurements actually cools the room. The fabrication hypothesis cannot explain this.

Cross-species convergence. A human, a bat, and a bee navigate around the same obstacle. Each uses a radically different mediational architecture – trichromatic vision, echolocation, compound-eye optics with ultraviolet sensitivity. Each converges on the same spatial constraint: there is something here that resists passage. The constraint-pattern is invariant across mediational systems that share almost nothing in their organisational structure. The convergence is strong evidence that the constraint is not an artefact of any particular mediation.

Cross-paradigmatic convergence. Newtonian mechanics and general relativity, operating within radically different conceptual frameworks, converge on gravitational predictions at ordinary scales. The conceptual mediations differ profoundly – absolute space versus curved spacetime, gravitational force versus geodesic motion – but the constraint-patterns they track converge within the domain of shared application. The convergence is evidence that both frameworks are responsive to the same underlying relational constraint, even though they describe it in incommensurable terms. A sceptic will note that this is precisely underdetermination: different descriptions fitting the same data. The reply is precise: convergence underdetermines ontology-of-things but supports ontology-of-constraints.[2] MEOW does not claim that Newton and Einstein describe the same objects. It claims they track the same invariant constraint-patterns. That is all the realism requires.

In each case, the structure of the argument is the same. Convergence across mediational variation is more parsimoniously explained by shared constraint than by coincidence. The more the mediational architectures diverge, the stronger the evidential force of convergence. And the convergence is never perfect – different mediations disclose different aspects of the constraint-pattern, at different resolutions, with different distortions. But the convergence is real, and it is what grounds MEOW's realism.

2 The underdetermination problem – that empirical evidence is in principle compatible with multiple incompatible theories – has been a persistent challenge to scientific realism since Quine (1951) and Duhem (1906). Stanford's (2006) 'problem of unconceived alternatives' sharpens the worry: the history of science shows that we have repeatedly failed to conceive of the very theories that would later replace our current best accounts. MEOW's response does not attempt to solve the underdetermination problem in its general form. It sidesteps it. The framework does not claim that convergence determines which theory is correct (underdetermination may indeed be irresolvable at that level). It claims that convergence across mediational variation is evidence for invariant constraint-patterns – structural features that any adequate theory in the domain must accommodate, regardless of its ontological commitments. Theories are underdetermined. Constraints are not – or at least, they are constrained by convergence in a way that theories-as-wholes are not. This is a weaker but more defensible realist commitment than the claim that our best theories describe reality as it is.

9.4 Objectivity as Convergent Stability

The convergence argument reframes objectivity.

On the traditional view, objectivity is the view from nowhere – the perspective that sees reality as it is, free from the distortions of any particular viewpoint. MEOW rejects this. There is no view from nowhere. Every encounter is mediated. Every perspective is a modal differentiation with its own salience structure, its own compression profile, its own ontological grammar.

But it does not follow that objectivity is impossible. It follows that objectivity is differently constituted than the tradition assumed.

Objectivity, within MEOW, is convergent stability across mediational variation. A claim is objective to the extent that the constraint-pattern it tracks is stable across encounters organised by different mediational architectures. 'Water boils at 100°C at sea level' is objective because the constraint-pattern it tracks – the phase transition of H_2O under specific pressure conditions – converges across thermometers, calorimeters, visual observation, and the practical experience of every culture that has boiled water. 'This painting is beautiful' is less objective – not because beauty is unreal, but because the constraint-pattern it tracks (aesthetic affordance relative to a particular cultural and perceptual formation) is less stable across mediational variation.

The gradient is continuous, not binary. Objectivity is not a property that claims either have or lack. It is a degree – a measure of how robustly the underlying constraint-pattern survives variation in the mediational architecture through which it is encountered. This maps directly onto the effectiveness gradient of Chapter 7: Invariants are highly objective (stable across radical mediational variation), Contestables are moderately objective (stable across limited variation), and Fluids are minimally objective (stable only within narrow mediational contexts).

This reframing preserves everything that matters about objectivity – the sense that some claims are more robustly grounded than others, that reality pushes back against bad descriptions, that not everything is equally valid – whilst discarding the metaphysical commitment that objectivity requires a God's-eye view. It requires convergence, not transcendence.

9.5 Against the Four Alternatives

MEOW's convergence realism occupies a position that is easily confused with several alternatives. The distinctions are worth drawing precisely, because the framework will be misidentified by readers who recognise familiar features and assume the whole package comes with them.

Naïve object realism holds that the world consists of independently existing objects with determinate properties, and that knowledge consists of accurately representing those objects and properties. MEOW rejects this. Objects are scale-dependent stabilisations, not substances. Properties are aspectual descriptions of encounter-organisation, not features of things-in-themselves. The convergence that grounds MEOW's realism is convergence on constraint-patterns, not on objects. The world does not consist of objects that we more or less accurately depict. It consists of relational constraint that we more or less stably navigate.

Scientific realism (in its standard form) holds that the entities posited by mature scientific theories exist approximately as described. MEOW is closer to scientific realism than to any other standard position, but it diverges on the ontological interpretation.[3] Science tracks constraint-patterns with extraordinary precision and cross-perspectival robustness. MEOW affirms this. But it does not interpret the resulting posits – electrons, genes, tectonic plates – as substances existing independently of all encounter. They are stable constraint-patterns within encounter-networks, disclosed through particular mediational architectures (instruments, experimental protocols, theoretical frameworks). Their reality is grounded in convergent stability, not in substance-ontological independence. The practical difference is minimal –

3 The standard scientific realist position is typically defended through the 'no-miracles argument' (Putnam, 1975; Boyd, 1983): the success of science would be a miracle if the theories that produce that success were not approximately true. MEOW's convergence realism offers a structural analogue of this argument: convergence across mediational variation would be a coincidence if the constraint-patterns disclosed were fabricated by the disclosing systems. The structure is parallel, but the conclusion differs. The no-miracles argument concludes that the entities posited by successful theories approximately exist. MEOW concludes that the constraint-patterns tracked by successful theories are approximately invariant. The distinction matters because it avoids the pessimistic meta-induction – the historical observation that past theories' entities have routinely been abandoned whilst their predictive structures have been retained. Convergence realism is resilient to the pessimistic meta-induction precisely because it commits to constraint-structure rather than to entity-catalogues.

scientists can continue doing exactly what they do. The philosophical difference is that the metaphysical interpretation no longer requires a commitment to substances behind the patterns.

Constructivism holds that the categories through which we understand reality are constituted by cognitive, cultural, or social practices. MEOW agrees that mediation shapes encounter-organisation. It disagrees that constraint is reducible to construction. The constructivist's error is to overstate the sovereignty of the mediational system.[4] Mediation shapes the form of encounter; constraint determines what is stable within it. An agent can construct any categories it likes. The constraint-field will cooperate with some and refuse others. The refusal is not constructed. It is encountered.

Kantian noumenalism posits an unknowable reality behind appearances – things-in-themselves that we can never access but must postulate as the ground of experience. MEOW rejects this. There is no hidden realm behind encounter. Constraint is internal to encounter-organisation, not a trace of something transcendent. The noumenal-phenomenal distinction, as Chapter 1 argued, is a post hoc abstraction imposed on encounter-events that were never bifurcated to begin with. MEOW does not posit unknowable things behind knowable appearances. It posits structured encounter in which constraint and mediation co-emerge, and in which invariance is inferred from convergence, not postulated as a hidden substrate.

Relativism holds that all claims are equally valid relative to their framework, and that no framework has privileged access to truth. MEOW rejects this. Constraint is not relative. Gravity does not vary with cultural perspective. Toxicity does not negotiate with ontological grammar. The convergence that grounds MEOW's realism is precisely the evidence that not everything

4 The critique of constructivism offered here should be distinguished from the standard realist critique, which typically accuses constructivism of denying the existence of an external world. MEOW does not make this accusation; it shares the constructivist's insight that mediation shapes encounter-organisation. The disagreement is narrower and more precise. Van Fraassen's (1980) constructive empiricism, for instance, holds that science aims at empirical adequacy rather than truth about unobservables. MEOW agrees that empirical adequacy is all that mediation can deliver – but adds that convergent empirical adequacy across heterogeneous mediational systems is evidence for something beyond mere adequacy. When radically different systems converge, the parsimony of the constraint-invariance hypothesis exceeds the parsimony of the coincidence-of-adequacy hypothesis. This is where MEOW parts company with van Fraassen: not on the limits of mediation, but on what convergence across those limits warrants.

is framework-relative. Some constraint-patterns persist across all mediational variation. Those patterns are what the framework calls invariant, and their invariance is what prevents MEOW from collapsing into relativism. The accusation of relativism, as noted in earlier chapters, reveals more about the accuser's assumptions than about the framework's commitments: it presupposes that the only alternative to substance-ontological realism is relativism, when MEOW offers a third option – realism grounded in convergent constraint rather than in mind-independent substance.

9.6 The Strength and Limits of Convergence

The convergence argument is strong but not unlimited. Intellectual honesty requires identifying where it holds firmly and where it reaches its limits.

Convergence is strongest where mediational diversity is greatest and constraint-patterns are most stable. The physical sciences occupy this territory. Fundamental constants, conservation laws, and well-confirmed causal regularities have been stabilised across instruments, species, cultures, and theoretical paradigms spanning centuries. The convergent stability of these constraint-patterns is the best evidence MEOW has – or that any realism has – for invariant structure.

Convergence weakens as constraint-patterns become more fluid, more scale-dependent, or more sensitive to mediational variation. Biological categories (species, gene, organism) exhibit moderate convergence – robust enough for scientific practice, but less stable across theoretical paradigms than physical constants. Social categories (justice, freedom, harm) exhibit weak convergence – heavily dependent on cultural mediation, ontological grammar, and institutional context. Phenomenal categories (qualia, aesthetic experience, emotional states) exhibit minimal convergence – highly sensitive to individual mediational variation and resistant to the synchronisation protocol's discretisation pressure.

This gradient of convergent strength is not a defect of the framework. It is a prediction. If realism is grounded in convergent stability, then the degree of realism warranted for any domain should track the degree of convergence achievable in that domain. Physics warrants strong realism. Biology warrants moderate realism. Social theory warrants a more cautious realism – one that acknowledges the real constraint-patterns at work in social life (incentives, violence, dependence, scarcity) whilst recognising that their articulation is more heavily shaped by mediational and grammatical factors than physical constraints. Weak convergence does not mean absence of constraint. It means the constraints are real but their codings are plastic and ontologically distorted by the encoding protocol. Phenomenology warrants the most cautious realism of all – one that affirms the reality of experience whilst acknowledging that its description is structurally limited by the encoding protocol's bandwidth.

This graduated realism is not relativism. It does not say that all domains are equally uncertain. It says that epistemic confidence should scale with convergent stability, which varies by domain. It preserves the realist commitment – constraint is real, invariance is evidentially grounded – whilst refusing to extend that commitment beyond what the evidence supports.

One important caveat: convergence can be manufactured by shared protocol constraints – identical instruments, identical training, identical operationalisation.[5] When every laboratory uses the same equipment, the same procedures, and the same theoretical framework, convergence may reflect methodological homogeneity rather than invariant structure. Strong convergence therefore requires genuine heterogeneity in mediation and cross-context success in intervention. The convergence that grounds realism is not agreement among identical systems. It is agreement among systems that share as little mediational architecture as possible whilst still tracking the same constraint-patterns.

5 The concern about manufactured convergence is a version of what Collins (1985) calls the 'experimenter's regress': the competence of an experimental apparatus is judged by whether it produces the right result, but the right result is known only through competent apparatus. Applied to convergence realism, the worry is that convergence across instruments may reflect shared calibration practices rather than shared constraint-patterns. MEOW's response – that strong convergence requires genuine heterogeneity in mediation – is the structural answer to this worry. Cross-species convergence (bats and humans navigating the same obstacle) cannot be explained by shared calibration. Cross-instrumental convergence is evidentially weaker where instruments share design principles, and stronger where they do not. The chapter's honesty about this limitation – that convergence must be heterogeneous to be evidentially powerful – is itself part of the realist defence: a framework that identifies the conditions under which its own evidence is weak is more trustworthy than one that does not.

9.7 Filtering Does Not Imply Fabrication: The Full Defence

The principle introduced in Chapter 2 can now be given its full defence, because the convergence argument provides the justification that the earlier statement lacked.

Mediation shapes encounter. Every perception, every concept, every measurement is structured by the mediational architecture through which it is organised. This is a commitment of the framework, not a concession.

But shaping is not manufacturing. The evidence is convergence. If mediation manufactured constraint, then differently mediated encounters would not converge. A system that fabricates its own constraint-patterns would have no reason to produce the same fabrications as a radically different system operating through a different mediational architecture. The fact that convergence does occur – that bats and humans navigate the same obstacles, that thermometers and infrared cameras agree on temperature, that Newtonian mechanics and general relativity converge at ordinary scales – is evidence that what is being encountered is not fabricated by the encountering system.

The argument is not deductive. It does not prove that mediation never fabricates. It provides abductive justification: convergence across mediational variation is best explained by shared constraint, not by coincidence of fabrication. The more the mediational architectures diverge, the less plausible the fabrication hypothesis becomes. At some point – when organisms with entirely different sensory systems, instruments built on different physical principles, and theories with incommensurable conceptual frameworks all converge on the same constraint-patterns – the fabrication hypothesis becomes explanatorily untenable.

This is the empirical grounding that convergence realism requires. Not proof. Not deduction. Not access to things-in-themselves. But strong abductive inference from convergence to invariance – the same kind of inference that grounds every empirical science, now made explicit as the philosophical foundation of the framework's realism.

9.8 Realism and the Architecture

The realist defence developed in this chapter is not an appendage to the framework. It is structurally integrated.

Ontology (Chapters 1–2): relational constraint is the invariant anchor. Objects are stabilisations; constraint-patterns are what the stabilisations track.

Subjecthood (Chapter 3): modal differentiation is real organisational structure within the constraint-field, not a projection or an illusion. The subject's perspective is encounter-relative, but the constraint-field it differentiates is not.

Logic (Chapter 4): coherence grammar is a stability condition extracted from the operating requirements of encounter. Its necessity is structural – grounded in the constraint-field's refusal to sustain incoherent stabilisations.

Epistemology (Chapter 5): knowledge is stabilisation under encounter pressure. The encounter-pressure that tests stabilisations is exerted by the constraint-field. Bad stabilisations collapse because the field refuses to cooperate with them.

Perception (Chapter 6): affordance gradients are encounter-constituted but constraint-grounded. Cross-species convergence on shared environmental features is perceptual evidence for invariant constraint.

Language (Chapter 7): the synchronisation protocol is finite-bandwidth and ontologically non-neutral. Its limitations are real – but the constraint-patterns it attempts to encode are not created by the encoding.

Social ontology (Chapter 8): second-order constraint systems are real because they alter interactional trajectories. Their fragility is real because the protocol on which they depend has structural limits. Neither the reality nor the fragility is constructed by the framework's description of them.

At every level, the realism is the same: constraint persists across mediational variation, and convergence is the evidence for that persistence. The framework does not assume realism. It earns it, through the convergence argument developed in this chapter and instantiated across every preceding one.

9.9 What This Chapter Has Established

MEOW is a realist framework. Its realism is grounded not in substance-ontological independence but in the convergent stability of relational constraint across distinct mediational architectures. When differently mediated encounter-systems converge on the same constraint-patterns, that convergence is evidence of invariant structure.

Objectivity is convergent stability across mediational variation, not the view from nowhere. It admits of degrees: strong in domains where constraint-patterns are highly stable across radical mediational diversity, weaker in domains where constraint is more sensitive to mediational and grammatical shaping. This graduated realism is not relativism. It is realism calibrated to evidence.

The framework has been distinguished from naïve object realism, scientific realism, constructivism, Kantian noumenalism, and relativism. It shares features with several of these traditions but is reducible to none.

The principle that filtering does not imply fabrication has been given its full abductive defence. Convergence across mediational variation is best explained by shared constraint, not by coincidence of fabrication. This is the empirical grounding of convergence realism.

The architecture is now defended. What remains is to demonstrate its explanatory purchase beyond pure metaphysics, and to identify honestly where its reach ends. The next chapter applies the framework to a single, carefully delimited case – the contemporary debate about AI and mind – as a controlled demonstration of what the architecture can do. The final chapter identifies what it cannot.

10

Apophatic Minds

Application

The Dao that can be spoken is not the eternal Dao.

— *Laozi, Dào Dé Jīng*

A map is not the territory it represents, but, if correct, it has a similar structure to the territory, which accounts for its usefulness.

— Alfred Korzybski

10.1 Introduction: Why This Case

The preceding nine chapters have constructed an architecture. Encounter-events are ontologically primary. Constraint is the invariant anchor. Subjects are modally differentiated process-clusters. Logic provides coherence grammar. Knowledge is stabilisation under encounter pressure. Perception structures the field into affordance gradients. Language synchronises stabilisations through a finite-bandwidth protocol with ontological biases. Institutions are second-order constraint systems that inherit the limitations of that protocol. Convergence realism grounds the whole structure.

The question now is whether the architecture does anything.

A framework that explains only itself is a curiosity. A framework that explains something the reader already recognises as genuinely puzzling is a tool. This chapter applies the architecture developed in Parts I–III to a single, carefully delimited case: the contemporary debate about AI and mind. The choice is not arbitrary. The AI-mind debate is a live philosophical dispute that exhibits, in unusually clear form, the structural pathologies that MEOW's architecture is designed to diagnose – substance-ontological reification, ontological grammar distortion, and the systematic overestimation of language's capacity to track the phenomena it purports to describe.

The chapter does not attempt to resolve the question of machine consciousness. It demonstrates how the framework dissolves the conceptual confusions that make the debate intractable.

10.2 The Apophatic Structure

There is a pattern in how contemporary discourse defines 'mind' in relation to artificial intelligence, and once it is identified it becomes difficult to unsee.

Language generation was once treated as evidence of intelligence. When large language models began generating fluent prose, language was reclassified as 'statistical pattern completion'. Reasoning was evidence of mind. When machines began solving structured problems, reasoning became 'probabilistic inference'. Creativity was the mark of a distinctive cognitive capacity. When AI produced novel recombinations across artistic media, creativity was downgraded to 'remix'.

The pattern is consistent:

1. A capacity is identified as evidence of mind.
2. A machine demonstrates that capacity.
3. The capacity is reclassified as mechanical.
4. The definition of mind retreats to a new, harder-to-demonstrate property.

The process repeats until what remains is a cluster of phenomena nobody can positively specify: consciousness, subjectivity, qualia, the what-it-is-like-ness of experience. This residue is treated as the essence of mind – the irreducible remainder that machines lack by definition.

None of this is new. The pattern has been noticed for decades. It keeps reappearing because it is rhetorically convenient and structurally invisible to those who inhabit it.[1]

It has a structure worth diagnosing. The structure is apophatic. The pattern described here is not every position in the philosophy of mind; functionalists, illusionists, and biological naturalists each resist it in different ways.[2] It is the dominant rhetorical equilibrium of popular and semi-tech-

1 The pattern of definitional retreat has been observed under various names. Turing (1950) anticipated it: his proposal of the 'imitation game' was partly an attempt to sidestep definitional disputes about intelligence by replacing the substance question ('can a machine think?') with a behavioural one. Dreyfus (1972, 1992) documented how AI optimists repeatedly redefined the goalposts when their claims about machine intelligence proved premature, though Dreyfus's diagnosis ran in the opposite direction – he argued that the goalposts were substantive and that machines would never reach them. More recently, the 'AI effect' has become a commonplace in popular science writing: as soon as AI achieves a task, it ceases to be considered AI. MEOW's contribution is not the observation of the pattern but its structural diagnosis: the retreat is sustained by the interaction of apophatic method, substance-ontological grammar, and the post hoc noumenal-phenomenal split. Without the diagnosis, the observation is merely anecdotal.

2 The qualification is important and deserves expansion. Functionalism (Putnam, 1967; Block, 1978) identifies mental states with functional roles rather than with substances, and thereby avoids the worst excesses of the apophatic retreat – a functionalist can in principle grant mentality to any system that instantiates the right functional organisation, regardless of substrate. Illusionism (Frankish, 2016) denies that phenomenal consciousness is what it seems, arguing that introspection systematically misrepresents the nature of experience; this dissolves the 'irreducible remainder' by denying that it names what its defenders think it names. Biological naturalism (Searle, 1992) roots consciousness in biological processes specifically, resisting both the apophatic retreat and the substrate-neutral extension. MEOW's diagnosis targets the discursive default of popular and semi-technical debate, not

nical discourse – the default mode in which the debate is publicly conducted.

In theology, the via negativa is the tradition of approaching God through successive negation.[3] When the object of inquiry is posited as real but resists positive description, discourse proceeds by saying what it is not. God is not finite, not temporal, not material, not changeable. What remains after the negations is treated as the essence – not because it has been positively identified, but because everything else has been stripped away.

Contemporary AI discourse has adopted the same epistemic regime. Mind is no longer defined by what it demonstrably does. It is defined by whatever machines demonstrably do not do – yet. Each successful demonstration of a cognitive capacity by a machine is absorbed into the 'not-mind' column, and the concept retreats further into the territory of the unspecifiable.

This is not merely analogous to apophatic theology. It is the same structural pattern. When a concept is posited as real but exceeds the descriptive capacity of available language, discourse becomes apophatic. The result is what might be called, without excessive charity, the mind of the gaps – a concept that shrinks whenever AI expands, whose content is defined entirely by what remains unautomated.

---

these technically precise positions. The default is more influential than the philosophical alternatives precisely because it is grammatically enforced and does not require philosophical sophistication to inhabit.

3 The apophatic tradition in Christian theology runs from Pseudo-Dionysius the Areopagite through Meister Eckhart, John of the Cross, and the anonymous author of The Cloud of Unknowing. Turner (1995) provides the most rigorous contemporary analysis, arguing that apophatic theology is not merely a method of negation but a recognition that the object of theological discourse exceeds the descriptive capacity of language. Bonhoeffer's (1953/1971) later writings push the logic further, toward a 'religionless Christianity' in which God is encountered in the world rather than posited beyond it. The structural parallel deployed in this chapter is not decorative. Turner's analysis of how negation produces the appearance of a hidden object – and how the hidden object is then invested with the significance of everything that was negated – maps precisely onto the dynamic identified in AI–mind discourse. The 'mind of the gaps' is structurally identical to the 'God of the gaps', and the mechanism is the same: apophatic method operating within a substance-ontological grammar.

10.3 The Double-Entry Ledger

The apophatic dynamic operates through a transfer mechanism that the architecture of this monograph makes legible.

The history of science and religion provides the structural precedent. Thunder was divine agency until meteorology claimed it. Epilepsy was demonic possession until neurology claimed it. The origin of species was divine craftsmanship until evolutionary biology claimed it. Each transfer follows the same logic: credit science, debit religion. The books balance, but only because the religion column keeps shrinking. What remains after centuries of transfers is an ineffable residue – a transcendent something that science supposedly cannot reach. God becomes whatever science has not yet explained.

AI discourse runs the same ledger with mind and mechanism. Credit mechanism, debit mind. Each time a machine demonstrates a capacity previously attributed to mind, that capacity migrates columns. What remains is the irreducible remainder – consciousness, subjectivity, phenomenal experience – which acquires increasing reverence precisely because it is all that is left.

The ledger does not prove mind is unreal. It makes mind conceptually homeless – a noun in search of predicates, sustained by the grammar that requires it and the reverence invested in whatever has not yet been transferred.

10.4 The Retreat into Experience

The apophatic retreat always terminates at the same point: subjective experience.

Whatever else is stripped away, this much is declared secure. There is something it is like to be a human – to see red, to feel pain, to taste coffee. This phenomenal character, this first-person interiority, is treated as the final line of defence. AI may reason, may speak, may create, but it does not experience. It lacks the inner light.

The claim is probably correct. On the organisational criteria developed in Chapters 1 and 3, contemporary LLM-style systems show no robust evidence of reflexively integrated encounter-organisation. The question is not whether the claim is true. The question is what metaphysical work the claim is being asked to do.

Notice what has happened. The concept of mind has been progressively emptied of every empirically demonstrable property, until what remains is the one phenomenon that is – by its nature – empirically inaccessible from outside the system that instantiates it. The retreat terminates at the point where verification becomes impossible. That may mark a real organisational threshold. What it does not warrant is treating 'experience' as a hidden commodity some systems possess, like contraband in a metaphysical suitcase. And it rests on a particular metaphysical commitment that the debate rarely examines: the assumption that subjective experience is a thing that a system possesses – a hidden ingredient, a metaphysical possession that some systems have and others lack, the way some boxes contain a jewel and others are empty.

This assumption is the substance-ontological commitment that MEOW's architecture dissolves.

10.5 The Reification Error

The framework developed in this monograph provides a precise diagnosis of what goes wrong when subjective experience is treated as a metaphysical possession.

Experience, within MEOW, is not a thing. It is a mode of encounter-organisation. Phenomenal presentation – the what-it-is-like-ness of experience – arises when mediational patterns achieve reflexive integration: when the encounter incorporates its own prior states, differentiates self from other, and integrates sensory, affective, and cognitive signals into a unified field of oriented engagement. This is a structural achievement, not a substance. It is something that certain encounter-organisations do, not something they have.

When defenders of human uniqueness invoke subjective experience as the final border between mind and mechanism, they commit a category error diagnosed by the framework at multiple levels:

They reify an organisational feature into a substance. The reflexive integration that produces phenomenal presentation is an organisational profile – a way that certain self-maintaining process-clusters structure their encounters. Treating it as a 'thing' that systems either possess or lack converts an organisational achievement into a metaphysical object. This is the same reification that Chapter 2 diagnosed in substance ontology generally: the conversion of stabilised patterns into self-subsisting entities.

They treat a structural distinction as a metaphysical border. The distinction between systems that exhibit reflexive integration and systems that do not is real and important. But it is a distinction in organisational profile, not a distinction between systems that contain a special substance and systems that do not. The border between 'has experience' and 'lacks experience' is being treated as a line between metaphysical categories, when it is better understood as a threshold in organisational complexity.

They mistake the remainder for a discovery. The apophatic procedure strips away every positively specifiable property and then treats what remains as the essence. But the remainder was produced by the stripping. It is the residue of the analytical method, not an independent finding. The 'irreducible remainder' of subjective experience is not irreducible because it names a fundamental feature of reality. It is irreducible because the method of successive negation guarantees a residue, and the residue is then invested with the significance that the negated properties can no longer carry.

10.6 The Grammatical Trap

The ontological grammar thesis developed in Chapter 7 provides a further layer of diagnosis.

The AI-mind debate is conducted in English – a noun-dominant, subject-predicate language that installs substance-ontological commitments at the level of syntax. Every sentence about mind is grammatically structured to presuppose a substance:

'The mind perceives.' 'AI lacks consciousness.' 'Humans possess subjective experience.'

Each sentence places a noun-subject in grammatical relation to a predicate. Each presupposes an entity – mind, consciousness, experience – that exists as a thing, prior to and independent of what it does. The grammar makes it almost impossible to talk about organisational profiles without converting them into substances in the act of speaking about them.

The apophatic structure diagnosed in Section 10.2 is partly a grammatical phenomenon. The via negativa is what happens when a language built for objects is forced to describe something that is not one. Each negation removes a predicate from the noun-subject 'mind', but the noun persists – grammatically required, syntactically load-bearing – even as its predicative content is progressively emptied. The grammar guarantees that 'mind' survives the negation process as a noun, even when every positive property has been stripped from it. Once 'mind' is a noun, the debate becomes a customs search: who has it, who lacks it, how to detect it, what substances it contains. The residue is not a metaphysical discovery. It is a grammatical shadow – the noun that the encoding protocol refuses to release.

This is not to say that the debate is merely grammatical. The organisational differences between human cognition and machine processing are real. The reflexive integration that produces phenomenal presentation is a genuine structural feature that current AI architectures lack. But the way the debate frames these differences – as the presence or absence of a substance called 'mind' – is shaped by the ontological grammar of the language in which the debate is conducted. A language that did not require noun-subjects, or that treated organisational profiles as grammatically primary, would not generate the same metaphysical picture.

10.7 The Noumenal-Phenomenal Distinction as Post Hoc

The deepest layer of the diagnosis draws on the argument developed across Chapters 1 and 2: the noumenal-phenomenal distinction is retrospective, not foundational.

The standard philosophical architecture treats the distinction between things-as-experienced and things-in-themselves as an a priori condition of experience. The split is assumed to be there before investigation begins. The apophatic method then operates within this architecture: remove the phenomenal properties, approach the noumenal core.

MEOW reframes the distinction as post hoc. Encounter-events are structurally unified. Mediation, presentational structure, and constraint co-emerge within a single relational organisation. The distinction between 'the experience' and 'the thing behind the experience' is imposed retrospectively – an analytical cut that carves a unitary event into aspects for the purposes of description. The cut is useful. It is not ontologically given.

If the noumenal-phenomenal distinction is retrospective rather than foundational, then the apophatic method is not uncovering a hidden object.

It is generating the appearance of one. Subtraction is not excavation. It is a way of producing emptiness whilst keeping the noun that names the thing you hoped was inside. The act of subtraction – remove language, remove reasoning, remove creativity – produces a remainder, and the remainder is mistaken for a discovery. The mysterious 'mind' at the centre of the debate is the shadow cast by the method, not the thing the method was searching for.

The defenders of the irreducible remainder are not wrong to point at something. The organisational features they gesture toward – reflexive integration, phenomenal presentation, the structural basis of subjective experience – are real. But the 'something' is an organisational profile, not a substance. And the conviction that it names a metaphysical essence is generated by the combination of apophatic method, substance-ontological grammar, and the post hoc noumenal-phenomenal split. Remove any one of these three, and the conviction loses its force.

10.8 The Mind as Heuristic Container

The framework's diagnosis converges on a precise claim about the concept of 'mind'.

Mind is not a natural kind. It is a heuristic container – a label that bundles together a cluster of organisationally diverse phenomena (consciousness, agency, intention, reflexivity, phenomenal experience, inferential capacity) under a single noun because human social life requires it.[4] 'Heuristic container' does not mean 'fiction'. It means 'compressed coordination handle over multiple real organisational profiles'. The bundling serves a prag-

4 The claim that 'mind' is not a natural kind but a coordination device aligns MEOW with a heterogeneous set of deflationists about mental categories, though the specific formulation is the framework's own. Dennett (1991) argues that consciousness is not a single phenomenon but a 'bag of tricks' – a collection of functional capacities that we bundle under a single term for practical reasons. Schwitzgebel (2016) has defended the view that there is no sharp boundary between the conscious and the non-conscious. The Buddhist philosophical tradition – particularly Nāgārjuna's Mūlamadhyamakakārikā as interpreted by Garfield (1995) – denies inherent existence to all categories including selfhood, treating them as conventional designations rather than markers of intrinsic nature. MEOW's 'heuristic container' is structurally closest to the Buddhist analysis: 'mind' is a conventional designation sustained by coordination pressure, not a natural kind discovered by metaphysical investigation. The framework's added contribution is the mechanism: ontological grammar explains why the conventional designation persists and how it acquires the appearance of naming a substance.

matic function: we need to track agency, intention, pain, and responsibility, because civilisation depends on doing so. 'Mind' is the compact label for that tracking apparatus.

The container persists across cultures not because it names a metaphysical substance but because it solves a perennial coordination problem. Stability of usage is evidence of recurrent coordination pressure, not of a metaphysical joint in nature.

The AI debate keeps asking: does this machine have a mind?

The framework suggests a more productive question: does the machine exhibit any of the specific organisational features that have been bundled into the heuristic container called 'mind'? And if so, which ones, and to what degree?

That is a gradient question, not a binary one. It replaces the substance question (does the machine possess the thing?) with a structural question (does the machine exhibit the organisational profile?). The structural question is answerable – not easily, but in principle – through analysis of the system's encounter-organisation. The substance question is not answerable, because there is no substance to find.

10.9 The Hard Problem Reframed

A brief note on the so-called Hard Problem of consciousness, which looms over this debate and which the framework reframes rather than resolves.

The Hard Problem, as formulated by Chalmers, asks: why does physical processing give rise to subjective experience?[5] Why is there 'something it is

5 Chalmers's (1996) formulation of the Hard Problem has shaped the philosophy of consciousness for three decades. The 'easy problems' – explaining cognitive functions, behavioural responses, and neural correlates – are hard in practice but straightforward in principle: they ask for mechanisms. The Hard Problem asks something different: why is there phenomenal experience at all? Why doesn't the mechanism run 'in the dark'? Levine's (1983) 'explanatory gap' makes a related point: even a complete physical account of consciousness would leave unexplained why those physical processes are accompanied by subjective experience. MEOW does not close either gap. What it does is relocate the question. The Chalmers–Levine gap presupposes a bifurcation between physical mechanism and phenomenal character – substance categories that MEOW replaces with encounter-organisation. The relocated question – why does reflexive integration produce phenomenal presentation? – is still hard, but it no longer has the shape of a substance-magic problem. It has the shape of an organisation-presentation problem, which is at least the kind of problem that structural analysis can address.

like' to be a conscious system, rather than nothing?

The problem presupposes the substance architecture that MEOW rejects. It assumes a bifurcation between physical processing (objective, third-person, mechanistic) and subjective experience (first-person, qualitative, phenomenal), and then asks how the first produces the second. Property dualism attempts to formulate the explanatory gap without substance commitments, distinguishing phenomenal from functional properties rather than mental from physical substances. Within MEOW, this distinction is interpreted as an aspectual differentiation within encounter-organisation rather than as an ontological divide – 'phenomenal' and 'functional' are classificatory stances applied to the same relational event, not ontological kinds. The question is generated by the noumenal-phenomenal split treated as foundational – whether in substance or property form – and MEOW rejects the split at both levels.

MEOW does not solve the Hard Problem. It changes the problem's shape. If experience is an organisational mode of encounter rather than a metaphysical add-on to physical processing, then the question is not 'how does matter produce mind?' but 'what organisational profiles produce phenomenal presentation?' The first question invites mysticism, because it asks how one substance generates a categorically different substance. The second question invites structural analysis, because it asks what conditions within encounter-organisation give rise to reflexive integration.

The mystery does not vanish. The framework does not dissolve the explanatory gap; it relocates it from substance-magic to organisation-presentation. It is entirely reasonable to wonder why reflexive integration produces phenomenal character rather than operating 'in the dark'. But the wonder is now directed at an organisational question – a question about the relationship between structure and presentation – rather than at a substance question about how inert matter secretes consciousness. The shift in framing does not eliminate the difficulty. It removes the theatrical architecture that makes the difficulty look like an impassable wall rather than a genuinely hard but structurally well-posed problem.

10.10 What This Chapter Has Demonstrated

The contemporary debate about AI and mind operates apophatically: mind is defined through successive negation of capacities shared with machines, producing an 'irreducible remainder' invested with the significance of everything that has been stripped away. This remainder – subjective experience – is then reified as a metaphysical possession, a substance that systems either have or lack.

The framework developed in this monograph diagnoses the debate at three levels. At the level of epistemic structure, the apophatic pattern generates a residue and mistakes it for a discovery. At the level of ontological commitment, the residue is reified – an organisational feature of certain encounter-systems is converted into a substance. At the level of conceptual genealogy, the noumenal-phenomenal distinction that underwrites the reification is itself a post hoc abstraction rather than a foundational given.

The ontological grammar of the language in which the debate is conducted reinforces all three errors: noun-dominant syntax preserves 'mind' as a grammatical subject even as its predicative content is progressively emptied, making the residue appear to name a thing rather than a structural shadow.

The chapter has not resolved the question of machine consciousness. It has shown that the question, as currently framed, is malformed – shaped by substance ontology, ontological grammar, and apophatic method rather than by the organisational features it purports to address. The productive replacement is a structural question about organisational profiles: what conditions produce reflexive integration, phenomenal presentation, and the other encounter-organisational features that the heuristic container 'mind' bundles together?

That question is answerable. It is just not answerable in the vocabulary the debate has been using.

The architecture has now been defended (Chapter 9) and demonstrated (this chapter). What remains is an honest accounting of where it reaches its limits. That is the subject of the final chapter.

11

Normativity

A Pseudo-Frontier and Its Dissolution

The opinion which is fated to be ultimately agreed to by all who investigate is what we mean by the truth.

— Charles S Peirce

A map is not the territory it represents, but, if correct, it has a similar structure to the territory, which accounts for its usefulness.

— Alfred Korzybski

11.1 What the Previous Chapter Promised

CHAPTER 10 CLOSED BY IDENTIFYING normativity as the architecture's principal open frontier. The description was accurate at the time: the framework could describe what constraints obtain, how coordination succeeds and fails, and how institutions stabilise and break. What it had not addressed was when constraint becomes obligation – when a pattern of enforced regularity acquires the character of something that ought to be maintained or resisted.

That framing was careful, but it was also provisional. It held the question open because the metaethical commitment required to close it had not yet been stated. This chapter states it, draws the consequences, and argues that what looked like a gap in the architecture is better understood as a pseudo-problem generated by a cognitivist assumption MEOW was never obliged to share.

The normativity frontier does not need closing. It needs dissolving. The procedure is the same one applied to the hard problem of consciousness in Chapter 10, to the realism-idealism impasse in Chapter 9, and to the no-relata demand in Chapter 2. In each case, the problem is not answered but traced – back to the structural misdrawing that produced it. Normativity is no different.

11.2 The Cognitivist Import

The is-ought gap – Hume's observation that no set of descriptive premises entails a normative conclusion without the addition of at least one normative premise – is a genuine logical observation. It is not in dispute here.[1] What is in dispute is the metaphysical gloss that moral cognitivism places upon it.

1 Hume's is-ought observation (Treatise, 3.1.1) is one of the most cited passages in moral philosophy, though its interpretation is contested. Some commentators read it as a strict logical barrier between descriptive and normative discourse; others read it as a more modest observation about the need to make normative premises explicit. The present chapter takes the stronger reading: no structural description of encounter-organisation entails a normative conclusion without a normative premise. This is accepted as a logical datum, not resisted. The chapter's argument is that this datum is only a problem for frameworks that expect their ontology to produce moral facts as output. A non-cognitivist framework has no such expectation. The is-ought gap is not a wall the architecture fails to scale. It is a wall the architecture has no reason to approach, because 'ought' is not the kind of thing structural description should be expected to deliver.

Cognitivism holds that moral claims are truth-apt: that 'you ought not to harm the innocent' is the kind of claim that can be true or false, that tracks something in the world, and that 'ought' names a property or relation whose existence is independent of our attitudes toward it. On this view, the is-ought gap represents an ontological barrier – a fundamental difference in the kind of thing described by descriptive and normative sentences, such that no amount of structural analysis of the world can produce moral facts as output.

It is this cognitivist framing that makes normativity look like a frontier for MEOW. If 'ought' names a mind-independent normative fact, and if MEOW's ontology of encounter-events, constraint, and mediation cannot produce such facts as outputs, then the framework appears to describe everything structurally relevant except what morality is supposedly about.

But that conditional is not one MEOW accepts. The cognitivist framing is not a pre-theoretical datum. It is a substantive metaethical commitment, and it is one the framework has no reason to share. Once the commitment is withdrawn, the gap changes character entirely.

11.3 The Non-Cognitivist Position

Moral non-cognitivism holds that normative utterances are not in the business of stating facts. They express attitudes, issue prescriptions, coordinate behaviour, ratify social norms, and stabilise shared expectations – but they do not describe a domain of moral reality that exists independently of the social practices in which they are embedded.

Moral non-cognitivism has several familiar forms – emotivist, prescriptivist, expressivist, quasi-realist – which differ in how they characterise the positive function of normative discourse. Those differences matter for metaethics. They do not matter for the present argument.[2] What matters here

2 The compressed enumeration deserves a brief map for the non-specialist. Emotivism (Ayer, 1936; Stevenson, 1944) treats moral utterances as expressions of emotion: 'stealing is wrong' expresses disapproval rather than stating a fact. Prescriptivism (Hare, 1952) treats them as universalisable imperatives. Expressivism (Gibbard, 1990) treats them as expressions of norm-acceptance. Blackburn's (1993, 1998) quasi-realism combines expressivism with a project of earning back the realist-seeming features of moral discourse (truth-aptness, objectivity, mind-independence) as pragmatic achievements rather than metaphysical discoveries. MEOW's position is closest to Blackburn's quasi-realism in that it treats moral discourse as socially real and functionally indispensable whilst denying that its apparent objectivity tracks mind-independent facts. The architecture's specific contribution is the mechanism: ontological grammar explains why moral discourse looks cognitivist – why

is the shared rejection of the cognitivist assumption: that 'ought' does not name a mind-independent normative property, that moral claims are not truth-apt in the robust descriptive sense presupposed by cognitivism,[3] and that the apparent objectivity of moral discourse is a functional consequence of its coordinative and legitimating role rather than evidence of a domain of normative facts.

This is the position the present framework adopts. It is stated here as a commitment, not derived from the architecture[4] – though it is, as the following sections argue, architecturally consonant with MEOW's other commitments in ways that are worth making explicit.

'duty' behaves grammatically like 'table' – and the internalisation account explains why it feels cognitivist. Quasi-realism earns back the surface features. MEOW explains why they were there to be earned back.

3 The most persistent technical objection to non-cognitivism is the Frege–Geach problem (Geach, 1960, 1965): if moral utterances are expressions of attitude rather than statements of fact, they should not be able to function as premises in logically valid arguments. But they do: 'if stealing is wrong, then getting your little brother to steal is wrong' has a valid conditional structure that requires the embedded 'stealing is wrong' to have the same content in the antecedent as it does when asserted independently. If the independent assertion merely expresses an attitude, what does the unasserted embedding express? This is a serious technical challenge, and the present chapter does not address it – not because it is unanswerable, but because a full response requires engagement with the formal semantics of normative vocabulary that lies outside the monograph's scope. A companion paper (Willis, in development) addresses the Frege–Geach problem within the ontological grammar framework. The short version: the problem arises because the logical form of embedded normative sentences presupposes that their content is propositional; the ontological grammar thesis suggests that this presupposition is itself a grammatical artefact of the encoding protocol rather than a constraint on what normative utterances can be.

4 The decision to adopt non-cognitivism as a commitment rather than derive it from the architecture is deliberate and bears comment. One could attempt to argue that MEOW's ontology entails non-cognitivism – that an encounter-event ontology with no mind-independent substances leaves no room for mind-independent moral facts. But this would prove too much: it would make the metaethical position a consequence of the ontology, and the ontology's vulnerability to revision would then destabilise the metaethics. The present chapter avoids this by treating non-cognitivism as an independent commitment that is consonant with the architecture rather than entailed by it. A reader who accepted MEOW's encounter-event ontology but preferred a cognitivist metaethics – perhaps a moral realism grounded in convergent constraint, analogous to the convergence realism of Chapter 9 – would disagree with this chapter without being forced to reject the architecture. The framework permits the disagreement. The chapter argues against it.

11.4 Why Non-Cognitivism Is Architecturally Consonant

Non-cognitivism does not follow from the encounter-event ontology, the account of ontological grammar, or the convergence realism defended in Chapter 9. It is an independent metaethical commitment. But it sits within the architecture without friction, and the architecture provides resources for understanding why moral discourse works the way it does – resources that cognitivism cannot easily supply.

Normative discourse as social coordination tooling

If encounter-events are ontologically primary, and if mind and world are derivative stabilisations within the encounter-field rather than pre-given substances, then moral discourse is most naturally understood as a specialised coordination mechanism – a region of the synchronisation protocol devoted to managing the shared navigational challenges that arise when multiple modally differentiated systems occupy overlapping affordance-fields.

Organisms with persistence conditions that depend on others – as social organisms by definition do – require mechanisms for managing conflict, stabilising cooperation, allocating resources, and sanctioning defection. Normative discourse is one such mechanism. It is not the only one: hierarchy, kinship, territory, and ritual are others. But it is the one that operates primarily through linguistic encoding, and it therefore inherits the structural features of the synchronisation protocol: its bandwidth limits, its ontological grammar, its capacity for false convergence, and its tendency to reify what it encodes.

On this account, moral language does not describe a pre-existing moral reality. It constitutes and maintains a social-coordinative reality – one that is real in exactly the sense that institutions are real: it constrains encounter, alters interactional trajectories, and produces consequences regardless of whether participants understand or endorse its underlying mechanics.

Internalisation and the phenomenology of obligation

A predictable objection: if normativity is 'merely' social coordination, why does it feel binding? Why does the agent who violates a norm experience not just the fear of sanction but the specific phenomenology of guilt, shame, and moral wrongness? Does the non-cognitivist account not simply change the subject – explaining coordination whilst leaving the felt obligatoriness of moral experience unexplained?

The architecture has a response, and it runs through the account of internalisation.

Second-order constraint systems, as Chapter 8 established, are not merely external enforcement mechanisms. They are transmitted through synchronisation, embedded in material and institutional structures, and – crucially – internalised through developmental exposure into the modal organisation of agents who grow up within them. A norm that begins as an external sanction-pattern can become, through repeated encounter, a structural feature of the agent's own affordance-organisation: a constraint on what feels possible, appropriate, and available. At that point, the norm is not experienced as an external imposition but as an internal necessity. The phenomenology of obligation is the felt texture of a successfully internalised constraint.

This is not a debunking account. It does not show that the experience of obligation is illusory or that its felt force is merely the residue of social conditioning to be transcended by rational reflection. It shows that the felt force is real – real as a feature of the agent's encounter-organisation – without requiring that it track a mind-independent normative fact. The experience is genuine. The metaphysical interpretation cognitivism places upon it is not.

Ontological grammar and the appearance of moral realism

The noun-dominant grammar of the synchronisation protocol does to normative discourse what it does to every other domain it encodes: it reifies. 'Duty', 'justice', 'rightness', 'obligation' are nouns. They name, in the grammar, stable entities with determinate properties. The grammatical form implies the existence of objects of moral perception – things that can be discovered, tracked, and known.

This grammatical reification is the primary source of moral cognitivism's apparent plausibility.[5] It is not that careful philosophers have examined

5 The claim that moral realism is a grammatical effect rather than a metaphysical discovery will be resisted by moral realists on the grounds that it is too deflationary – that it explains away the data rather than explaining it. The strongest version of this objection comes from moral naturalists (Boyd, 1988; Sturgeon, 1985) who argue that moral properties are natural properties and that moral facts are as robustly real as any other natural facts. On this view, 'cruelty is wrong' tracks a natural property (cruelty's tendency to diminish flourishing) and is true or false in exactly the way that 'water boils at 100°C' is. MEOW's response is that moral

the structure of moral experience and discovered that it reports mind-independent facts. It is that the grammar in which moral experience is encoded makes it look as though it does. The noun 'duty' behaves grammatically like the noun 'table'. It takes articles, admits of quantification, and functions as the subject of sentences with predicates that attribute properties to it. The grammar suggests an object. The suggestion is mistaken, but it is structurally generated rather than philosophically optional.

Moral cognitivism is, on this analysis, another instance of the reification pattern the architecture has diagnosed throughout – in the case of minds (Chapter 10), objects (Chapter 2), subjects (Chapter 3), and time. The apparent objectivity of moral facts is what stabilised coordinative practices look like once they have been encoded in noun-dominant grammar and internalised into the modal organisation of agents who experience their demands as given rather than constructed.

11.5 What Normativity Actually Is, Within MEOW

The positive account can now be stated plainly.

Normative discourse is a region of the synchronisation protocol specialised for the coordination, stabilisation, and legitimation of social constraint. It operates through expressive, prescriptive, and commissive speech acts rather than through descriptive ones. Its claims are not truth-apt in the cognitivist sense, but they are not merely subjective either: they express, stabilise, and enforce constraint-patterns that are socially real in the sense that matters – they alter what agents can do, must do, and are prevented from doing.

Normative utterances do not report mind-independent moral facts. They express attitudes shaped by internalised constraint-patterns, issue prescriptions that coordinate behaviour across agents, ratify social norms that sustain institutional second-order constraints, and stabilise shared orientations within overlapping affordance-fields. The 'ought' in a normative utterance is not reaching for a fact that 'is' cannot supply. It is performing a social act that 'is' does not perform – and performing it with a grammatical form that makes it look, misleadingly, as though it describes something.

naturalism conflates constraint-pattern tracking with moral fact discovery. The architecture can accommodate the observation that cruelty reliably disrupts organismal persistence and social coordination – that is a structural claim about constraint-patterns. What it does not accommodate is the further claim that this disruption constitutes a moral fact – an addition that requires the cognitivist assumption the chapter rejects.

Normativity may be socially indispensable without being ontologically primitive. This distinction is the load-bearing claim of the chapter. Social organisms require normative vocabularies. Agents embedded in institutions require normative frameworks for coordination, for legitimation, for the management of conflict. The practical necessity of normativity is not in question. Its metaphysical status is.

The societal need for normativity remains a live practical question. Its ontological status does not.

11.6 The Grammar Critique Has Normative Freight – But Not Moral Freight

A careful reader will press a further question. The architecture's grammar critique is itself a normative enterprise in a thin sense: it distinguishes better from worse ontological framings, identifies certain conceptual structures as systematically misleading, and recommends their replacement. Does this not smuggle a normative commitment through the back door – if not moral normativity, then at least epistemic normativity?

The answer is yes – but the concession does not threaten the non-cognitivist position. Epistemic normativity – the normativity of better and worse reasoning, clearer and less clear conceptual organisation, more and less adequate structural description – is not the same as moral normativity. The claim that substance ontology is less adequate than encounter-ontology is a claim about explanatory power, internal coherence, and comparative structural purchase. It is not a claim about what agents are morally required to do.

The grammar critique does carry one kind of normative implication that is worth making explicit: it can distinguish between normative frameworks that are epistemically honest about their social-coordinative character and those that are not. A normative framework that presents itself as tracking mind-independent moral facts, when the architecture's analysis suggests it is doing something quite different, is engaged in a form of ontological misrepresentation – not a moral failing, but a structural one. The framework can identify this. It cannot, on non-cognitivist grounds, derive from this identification that such frameworks are morally wrong to hold. It can say they are confused. That is a different and, in most cases, a sufficient criticism.

11.7 What the Dissolution Achieves

The dissolution of the normativity frontier does not leave moral philosophy without resources. It leaves it without a specific kind of spurious resource: the fiction of a domain of mind-independent normative facts that philosophical analysis might discover.

What remains after the dissolution is substantial. The architecture can explain how normative frameworks arise through coordination pressure, how they are stabilised through second-order constraint systems, how they are transmitted through synchronisation protocols, how they are internalised into modal organisation, and how they are reified by ontological grammar into the appearance of objective moral facts. This is a rich explanatory account of why moral discourse works as it does, why it generates persistent disagreement in the Contestables region of the effectiveness gradient, why moral frameworks function as powerful coordination tools and equally powerful instruments of domination, and why the felt authority of moral obligation is genuine without requiring a domain of moral facts to underwrite it.

It also provides the criterion mentioned in Chapter 8's closing pages: a MEOW-internal distinction between normative frameworks that track genuine structural constraint-patterns and those that are primarily enforcement fictions dressed in moral vocabulary. A normative framework organised around constraints that are structurally grounded in the persistence conditions and affordance-organisations of those it governs – one that responds to genuine coordination needs rather than serving the stabilisation interests of a subset of agents at the expense of others – is distinguishable from one that is not. The distinction is structural and diagnostic, not moral; it identifies explanatory and coordinative differences, not obligations written into the world. But it is not nothing. It is what an architecturally honest normative theory would have to acknowledge.

This book does not write that theory. It clears the ground on which it could be written without self-deception.

11.8 What This Chapter Has Established

The normativity frontier, as identified in the closing pages of earlier chapters, appeared open because the framework had not yet stated its metaethical commitment. With that commitment stated – moral non-cognitivism, consistently applied – the frontier dissolves rather than closes.

The is-ought gap is a genuine logical observation. The ontological barrier cognitivism constructs upon it is not. Once 'ought' is understood as expressive-coordinative social tooling rather than as a candidate description of mind-independent moral facts, the architecture's inability to derive normative conclusions from structural description is not a gap. It is the correct result.

Normative discourse is socially real. Its claims constrain encounter, alter interactional trajectories, and generate consequences with the same structural force as other second-order constraint systems. The reality of norms is not in question. Their ontological fundamentality is.

The phenomenology of obligation – the felt bindingness of moral experience – is explained in structural terms by the account of internalisation developed in this framework, without appeal to normative facts. It does not require normative facts as its object. It requires only that social constraint-patterns be sufficiently embedded in the modal organisation of agents that they are experienced as internal necessities rather than external impositions. That is a structural claim about the depth of internalisation. It is not a concession to cognitivism.

The grammar critique identifies ontological grammar as the mechanism through which coordinative practices acquire the appearance of moral objectivity. This is the same mechanism that produces substance-objects, persisting subjects, mind-world bifurcation, and the hard problem of consciousness. Moral realism is, from within this architecture, the characteristic product of ontological grammar applied to normative discourse. It is not a discovery. It is a grammatical effect.

The architecture does not generate moral prescriptions. It makes normative discourse auditable: it can identify when a normative framework is tracking genuine structural constraint and when it is performing the ideological function of stabilising the interests of some agents through the coordination machinery that governs all of them. That audit is not a moral judgement. It is a structural one.

That, for a framework whose method is diagnostic rather than prescriptive, is the appropriate place to stop.

The book that follows this chapter is the book the architecture has been building toward: not a completed normative theory, but an ontological framework honest enough to acknowledge what such a theory would require, and clear-eyed enough to have stopped short of pretending to supply it.

12

Conclusion

Architecture, Not Revelation

Our knowledge can only be finite, while our ignorance must necessarily be infinite.

— Karl Popper, *Objective Knowledge: An Evolutionary Approach*

We shall not cease from exploration,
And the end of all our exploring
Will be to arrive where we started
And know the place for the first time.

— T S Eliot, Little Gidding, *Four Quartets*

THIS MONOGRAPH HAS NOT delivered a final metaphysical picture. It has not revealed the hidden structure of reality, uncovered the essence of mind, or produced a theory of everything. It has done something more modest and, if the architecture holds, more useful: it has cleared conceptual space.

The space that has been cleared is the space occupied by substance ontology – by the assumption that reality consists of independently existing things bearing properties and entering into relations, and that the task of philosophy is to identify those things, describe those properties, and map those relations. That assumption organises the realism-idealism dispute, the mind-world bifurcation, the Hard Problem of consciousness, the no-relata demand, and the AI-mind debate. In each case, the architecture developed here has argued that the assumption generates the problem it purports to address. Remove the assumption, and the problem does not vanish – but it changes shape, often into something tractable rather than theatrical.

What has been put in its place is not a rival substance. It is an architecture – a set of structural commitments about how reality is organised, how agents are constituted within it, how knowledge is stabilised, how language synchronises and distorts, how institutions coordinate and fail, and where the limits of these processes lie.

The commitments can be stated briefly, because the monograph has stated them at length.

Reality consists of structured interactional processes. Encounter-events are ontologically primary. Mediation is constitutive, not obstructive. Constraint is the invariant anchor – patterned stability that persists across mediational variation – and convergence across distinct mediational architectures is the evidence for that invariance. Objects are scale-dependent stabilisations. Subjects are modally differentiated process-clusters whose self-maintaining organisation generates perspective, salience, and agency. Logic is a coherence grammar extracted from the operating conditions of stabilisation under constraint. Knowledge is stabilisation under encounter pressure, tested not by correspondence but by the persistent refusal of the constraint-field to cooperate with bad compressions. Perception structures the field into affordance gradients that are encounter-constituted and constraint-grounded. Language is a finite-bandwidth synchronisation protocol that aligns stabilisations across agents whilst imposing ontological commitments through its grammatical structure. Institutions are second-order constraint systems that inherit the limitations of the protocol through which they are constituted. And normativity – the question of when constraint becomes obligation – remains an open frontier, honestly identified.

Something has happened across these chapters that is worth naming, because it is the monograph's real intellectual move and it is easy to miss.

Four domains that philosophy typically treats as separate – ontology, epistemology, logic, and philosophy of language – have been unified under a single structural principle: constraint management within mediated encounter. Objects, knowledge, inference, and linguistic coordination are all treated as aspects of the same relational architecture, differentiated by scale and function but grounded in the same primitive. This unification was not announced as a programme. It emerged from the architecture. The framework does not claim to unify these domains. It does unify them, and the reader can assess whether the unification is forced or natural.

The unification has a further consequence. The Language Insufficiency Hypothesis, the ontological grammar thesis, the account of institutional false convergence, and the apophatic diagnosis of the AI-mind debate are not independent theses bolted onto a metaphysical core. They are structural consequences of the architecture – applications that fall out of the framework's commitments rather than being imported from outside. If the architecture is sound, the applications follow. If the applications fail, the architecture is under pressure. That is as it should be. A framework that cannot be tested by its consequences is not an architecture. It is a decoration.

The monograph has been written in a language that resists its own conclusions. Every sentence about relational process has been grammatically structured as a sentence about substances bearing properties. Every claim about encounter-organisation has been articulated through a noun-dominant protocol that converts processes into objects in the act of describing them. The fluency tax identified in Chapter 7 has been paid on every page. The reader has been asked to think against the grain of the encoding medium – to hold the structural content of the claims separately from the substance-ontological form in which they have been, unavoidably, expressed.

This is not an apology. It is a final instance of the ontological grammar thesis. The difficulty of articulating a non-substance ontology in a substance-grammar language is not a defect of the ontology. It is evidence for the thesis that grammatical structure shapes what can be fluently thought and expressed. The awkwardness is diagnostic. The cost of structural fidelity is articulatory resistance. The alternative – ease purchased through distortion – is what the monograph has spent eleven chapters diagnosing.

The architecture is offered as an open system, not a closed one.

Open, because it identifies its limits honestly. The normativity frontier is real. The formal relationship between bandwidth degradation and ontological mismatch distortion remains unspecified. The substrate-neutrality of modal differentiation is a principled commitment, not an empirical finding. These are not concessions but structural features of a framework that prefers honesty about scope to the illusion of completeness.

Open, because it is designed to ground further work rather than to conclude it. The framework developed here is the central spine of a broader intellectual project. The Language Insufficiency Hypothesis, already published independently, is a structural consequence of the architecture's account of language. The ontological grammar thesis extends into institutional analysis, governance critique, and the study of cross-cultural communication. The apophatic diagnosis of the AI-mind debate connects to broader questions about how discourse operates when its categories are outpaced by the phenomena they govern. Each of these lines of work presupposes the architecture without being contained by it.

Open, finally, because the architecture can be wrong. If the convergence argument fails – if cross-perspectival stability can be explained without invariant constraint – then the realist anchor dissolves and the framework loses its grounding. If the ontological grammar thesis overstates the influence of grammatical structure on metaphysical commitment – if substance thinking persists for reasons the grammar cannot explain – then the diagnostic apparatus is weakened. If the account of subjecthood as modal differentiation cannot accommodate the full range of phenomenal experience – if reflexive integration is insufficient as a structural criterion – then the framework's account of agents is incomplete. These are real vulnerabilities, not rhetorical gestures of humility. The architecture invites testing because it is designed to be tested.

The architecture developed here opens several lines of investigation that the present monograph does not pursue.

One concerns the formal relationship between bandwidth limitation, grammatical compression, and ontological distortion – a problem that connects the Language Insufficiency Hypothesis with the study of epistemic failure in large-scale institutions. The interaction between these factors determines the shape of the effectiveness gradient, and its precise characterisation remains an open formal question with consequences for governance, cross-cultural coordination, and the design of institutional language.

A second concerns the structural conditions under which reflexively integrated encounter-organisations arise – a question that bears directly on debates about artificial systems, biological cognition, and the emergence of phenomenal presentation. The framework provides the organisational criteria; the empirical investigation of which systems meet those criteria is work that the architecture enables but does not itself conduct.

A third concerns the unresolved normativity frontier identified in Chapter 11: whether normative structure can be derived from encounter-organisation itself, must be supplemented by additional commitments, or requires a reconceptualisation of the is-ought distinction as a grammatical artefact rather than an ontological given. Each of these paths is open, and each would extend the architecture into territory it has deliberately declined to enter prematurely.

Beyond MEOW does not reveal what reality is. It articulates the structures through which reality is disclosed – the encounter-organisations, constraint-patterns, synchronisation mechanisms, and institutional formations through which agents navigate a world they did not create and cannot fully describe. The world does not yield to our descriptions. It yields to our engagements. The descriptions are tools – lossy, grammatically biased, bandwidth-limited tools – that serve the engagements without replacing them.

The architecture is here. It is open. It is testable. And it is, by design, incomplete – because a framework that claimed otherwise would be revealing less about reality than about its own ambitions.

What This Book Claims, and How Firmly

Carlo Rovelli ends *Seven Brief Lessons on Physics* (2016) by sorting what he trusts from what he speculates. It is a generous gesture and a structurally honest one. This book attempts something similar, though with rather more architecture to account for.

The sorting is not modesty for its own sake. It is a prophylactic against a specific critical manoeuvre: the one that treats all claims in a framework as jointly liable, so that a speculative application can be used to impugn a structural commitment, or a contested empirical premise can be used to dismiss a diagnostic observation. That manoeuvre is not argument. It is category confusion with polemical intent. The following is an attempt to make it harder to perform.

What the framework takes from established science

The framework inherits several empirical commitments and does not pretend to have established them independently.

From physics: the relational and perspectival character of measurement, the thermodynamic asymmetry that gives temporal experience its directional structure, the absence of a privileged observational position. These are not philosophical glosses on physics. They are physics. The framework builds on them, but it did not build them.

From cognitive science and ecological psychology: the mediated character of perception, the constructive nature of attention, the bounded and compressive operations of finite cognition, the affordance-structure of the perceptual field. Gibson's ecological approach, the predictive processing literature, and the enactivist tradition are not allies adopted for rhetorical convenience. They are substantive sources for claims about how encounter-events are structured from the organismal side.

From evolutionary biology: the constraint that any account of subjecthood must accommodate the biological basis of modal differentiation — that organisms are not perspective-takers by convention but by evolutionary pressure. This is an empirical commitment, not a philosophical preference.

Any revision in these established fields may require corresponding revision in the architecture. That is not a weakness. It is what honest dependency looks like.

What the framework proposes as its own

The following are the framework's architectural claims. They are not derived from the sciences listed above, though they are consistent with them. They are philosophical proposals, and they are to be evaluated as such.

The ontological primacy of encounter-events. The claim that the encounter-event — structured, mediated, constraint-governed — is the appropriate primitive for ontology, and that mind and world are derivative abstractions stabilised within it rather than pre-given substances that encounter each other from the outside. This is the load-bearing commitment of the whole architecture. Everything else is downstream of it.

Ontological grammar as a mechanism of reification. The claim that natural languages carry a grammatical structure — noun-dominant, substance-implying, agency-distributing — that systematically encourages the treatment of process-stabilisations as things, and heuristic categories as natural kinds. This is not a claim about language in general. It is a claim about a specific and diagnosable feature of the linguistic encoding protocol that philosophy has been running on, largely without inspection.

Modal differentiation as the structure of subjecthood. The replacement of the 'subject as coordinate' model with the 'subject as modal organisation' model: the claim that what we call a subject is a pattern of differential salience and affordance-responsiveness, not a point of view on the world from the outside.

Coherence zones as heuristic stabilisations. The claim that the apparent unities we encounter — objects, persons, institutions, concepts — are not natural kinds but stabilised compressions that serve navigational purposes and may not survive structural scrutiny. This has diagnostic consequences wherever a philosophical problem turns on the assumption that a category is carving reality at its joints.

Convergence realism. The claim that the realist anchor is provided not by access to mind-independent substance but by cross-perspectival convergence under constraint — the progressive elimination of perspective-specific variance as a criterion for structural invariance. This is a positive proposal, not merely a rejection of nave realism.

These claims are defended in the relevant chapters. The defence is philosophical: by explanatory purchase, internal coherence, and comparative adequacy against the frameworks they replace. The reader who finds them unpersuasive is invited to say which framework does the same work more economically.

What the framework offers as diagnosis

The following are applications of the architecture to inherited philosophical problems. Their claim to acceptance is pragmatic: they dissolve or reframe problems that have proved intractable under prior treatment. They do not merely assert that those problems were confused. They show, in structural detail, how the confusion was produced.

The hard problem of consciousness. Reframed as a product of the nominalisation of 'experience' into a substance, followed by the demand that this substance be located causally within a physical system. The problem is not discovered. It is constructed by the analytical procedure. The architecture does not solve the hard problem. It explains why treating it as a problem was a mistake — which is a different and, arguably, more useful service.

The realism-idealism impasse. Reframed as a consequence of the shared substrate assumption: both realism and idealism accept the subject-object architecture and quarrel about which side has ontological priority. The framework rejects the substrate. The impasse is an artefact of the terrain, not a feature of reality.

The mind-of-the-gaps dynamic in AI discourse. Chapter 10's diagnosis of how 'mind' functions apophatically — acquiring its content by successive negation of capacities machines come to exhibit — is a direct application of the reification and heuristic-compression accounts. Whether this diagnosis is accepted will depend partly on whether the reader accepts the architectural claims above. That is as it should be.

Institutional ontology. The treatment of institutions as second-order constraint systems — protocols that govern the production of first-order protocols — is a diagnostic application of the social ontology framework. It is testable by its explanatory purchase on institutional behaviour, including failure modes.

What remains open

Intellectual honesty requires naming the frontiers, not only the achievements. The following are genuine open questions, not rhetorical gestures at humility.

Normativity. The architecture describes constraint, affordance, and coordination. It does not derive obligation. Chapter 11 identifies three candidate entry points — organismal interests, coordination pressure, and institutional enforcement — and explains why each is necessary but insufficient. The normativity frontier is real. It is the framework's most significant open border. The suggestion that the is-ought gap may itself be a grammatical artefact rather than an ontological given is noted, and it is the most architecturally consonant of the available routes forward. It has not been taken.

Formal modelling. Several structural claims — the effectiveness-complexity gradient, the modal differentiation account, the convergence criterion for realism — invite formalisation that this monograph does not attempt. Whether the claims survive rigorous formalisation is an open question. The architecture is not committed to their failing it. It is simply not the work of this book.

Empirical tractability. The framework identifies what would count as evidence for several of its claims, but does not pursue the empirical programmes those identifications imply. This is a limit of scope, not a structural deficiency. Whether the empirical investigations, if pursued, would vindicate or revise the architecture remains to be established.

The normativity frontier aside, the open questions listed here are second-order: they concern the elaboration and testing of commitments already made, not the commitments themselves. The reader who thinks this distinction is self-serving is invited to make the argument. The reader who accepts it is in a position to evaluate the architecture on its own terms.

A note on confidence

This book is confident about some things. The encounter-event primitive, the diagnosis of substance ontology as the primary source of philosophical pseudo-problems, the account of ontological grammar as a vehicle for reification — these are held with conviction. The confidence is not dogmatic. It is the product of sustained engagement with the alternatives and the judgement that they do not do the work as well.

The book is more provisional about the diagnostic applications, more speculative about the normativity frontier, and explicitly agnostic about the outcomes of empirical programmes it identifies but does not pursue.

Rovelli sorted his claims into what he knows, what he thinks, and what he dreams. The present taxonomy is somewhat less poetic: what the framework inherits, what it proposes, what it diagnoses, and what it leaves open. The categories are drier. The epistemic situation they describe is roughly the same.

Philosophy that cannot sort its claims by confidence level is not humble. It is evasive. This book has tried to be neither.

APPENDIX

A

Formal Summary of Commitments

A condensed restatement for reference and engagement

The following is a numbered summary of the framework's core commitments as developed across the monograph. It is intended as a reference tool – a compressed statement of what the architecture claims, for readers who wish to engage with the commitments directly, test them against objections, or compare them with adjacent positions. The summary does not replace the arguments. It indexes them.

I. Ontological Commitments

A1. Ontological Primacy of Encounter-Events. All that exists exists through encounter-events: structured relational occurrences in which mediation, presentational structure, and constraint co-emerge. Encounter-events are the ontological primitive. Relata are abstractions over stable patterns within such events, not independently existing entities.

A2. Relational Primacy. Relations are ontologically primitive. Relata – objects, subjects, properties – are derived: they are stable configurations abstracted from patterns within relational organisation. The demand for relata behind relations presupposes the substance ontology the framework rejects.

A3. Mediation as Constitutive Structure. Every encounter-event is organised through mediation – biological, cognitive, and linguistic-cultural. Mediation is not a veil between mind and world but the structural condition under which encounter becomes possible. No access occurs outside mediation.

A4. Constraint as Intrinsic Stability. Constraint is disclosed within encounter-events but is not reducible to mediation within them. It is the dimension of stability, resistance, or invariance that is not freely alterable by variation in mediation. Constraint is not projection (idealism) and not a noumenal intrusion (Kantian realism). It is internal to the relational structure of the encounter-event itself.

A5. Co-Emergence of Mediation, Presentational Structure, and Constraint. Mediation, presentational structure, and constraint are not sequential components but aspectual descriptions of a single relational organisation. No aspect is metaphysically prior; all arise jointly in the event's organisation.

A6. Derivative Status of 'Mind' and 'World'. 'Mind' and 'world' designate stable patterns abstracted from regularities across encounter-events rather than independently existing substances. Neither term denotes a domain external to encounter.

A7. Rejection of the Mental-Physical Dichotomy. 'Mental' and 'physical' are aspectual predicates applied to different emphases within encounter-event organisation. They identify classificatory stances, not ontological kinds.

A8. Relational Closure. All experiential, epistemic, and causal processes occur within networks of encounter-events. There is no need – and no conceptual room – for positing extra-encounter substances or forces.

A9. Existence and Disclosure. Existence consists in invariant constraint-patterns. Disclosure is their articulation within mediation. The constraint-pattern does not depend on encounter for its persistence. It depends on encounter for its manifestation.

II. Objects and Scale

A10. Objects as Scale-Dependent Stabilisations. Objects are not ontological furniture but scale-dependent stabilisations of interactional processes: dense, slow-changing constraint clusters relative to the organismal morphology and energetic structure of the encountering system.

A11. Durative Priority. Stability is slow interaction. Identity is stabilised duration. Substance ontology treats persistence as primitive and change as derivative. The framework reverses this priority: becoming precedes being, and persistence is derivative of stable constraint-transmission.

A12. Filtering Does Not Imply Fabrication. Mediation shapes how an encounter is organised but does not manufacture the constraint-patterns that persist across mediational variation. Shaping is not manufacturing. Convergence across distinct mediational architectures is evidence for invariant constraint, not coincidence of fabrication.

III. Subjecthood

A13. Subjecthood as Modal Differentiation. Subjecthood is the modal locus at which the interactional field becomes differentiated into constraint and affordance relative to a self-maintaining, constraint-responsive process cluster. The subject is not outside the field; it is a dynamic differentiation within it.

A14. Substrate Neutrality. Biological organisms are the paradigmatic but not the sole possible realisation of modal differentiation. The organisational profile, rather than the material substrate, determines whether modal differentiation obtains.

A15. Conditions for Subjecthood. Three conditions distinguish subjecthood from mere self-maintenance: self-maintenance must be intrinsic (the system bears the cost); constraint-responsiveness must be integrated (not externally bolted on); and the system must exhibit a unitary salience profile (a coherent centre of valuation).

A16. Subjecthood Requires Reflexive Integration. A subject is a modally differentiated cluster that exhibits recursive modelling of its own persistence conditions as an input to action selection across counterfactual affordance space.

A17. The 'I' as Indexical Compression. The first-person pronoun compresses the full modal differentiation of a self-maintaining process cluster into a single referential token. The compression is linguistically indispensable and ontologically misleading.

A18. The 'Self' as Diachronic Narrative Compression. The self is a stabilised pattern tracking organisational continuity across temporally extended encounter-sequences. It is a heuristic – real in what it tracks, misleading when mistaken for a substance.

A19. Agency Without Substance. Agency is the capacity of a modally differentiated system to alter its own encounter-trajectory in response to its salience structure. It does not require a Cartesian ego or an uncaused cause.

IV. Logic and Epistemology

A20. Logic as Coherence Grammar. Logic is the minimal structural condition required to preserve consistency in stabilised orientation under constraint. It is neither Platonic (existing independently of all encounter) nor conventional (adopted by agreement). It is a stability condition extracted from the operating requirements of encounter.

A21. Domain-Sensitivity of Logical Grammar. Classical subject-predicate logic is the coherence grammar appropriate to well-discretised, object-shaped domains. In domains where constraint-patterns are processual, relational, or graded, alternative logics provide the appropriate coherence grammar. The proliferation of logical systems is evidence of domain-sensitivity, not arbitrariness.

A22. Knowledge as Stabilisation Under Encounter Pressure. Knowledge is not the possession of internal representations corresponding to an external reality. It is the capacity of a modally differentiated system to maintain effective stabilisations under the ongoing pressure of encounter. The test of knowledge is resilience under constraint, not correspondence.

A23. Error as Structural. Error arises from compression loss, modal divergence, and ontological grammar distortion. It is a structural feature of finite-bandwidth mediation, not a contingent failure of intelligence.

A24. Objectivity as Convergent Stability. Objectivity is convergent stability across mediational variation. A claim is objective to the extent that the constraint-pattern it tracks is stable across encounters organised by different mediational architectures. Objectivity admits of degrees determined by convergent stability.

V. Perception and Affordances

A25. Affordances as Encounter-Constituted Gradients. Affordances are organism-field compatibility gradients constituted by encounter. Constraint topology is encounter-independent structure; affordance is that structure rendered navigable relative to a particular organisational profile. Affordances are relational, continuous, and context-sensitive.

A26. Meaning as Salience-Weighted Affordance Structure. Meaning is salience-weighted affordance structure relative to organismal persistence. For humans, persistence conditions include social continuity, role-stability, and symbolic commitments. Meaning is pre-linguistic: it arises in encounter before linguistic encoding.

A27. Perception as Orientation, Not Representation. Perception is the ongoing structured orientation of a modally differentiated system within its affordance gradients. It is a continuous process of adjustment, not the construction of inner pictures from outer data.

A28. Presentational Structure and Phenomenal Presentation. All encounter-events possess presentational structure (an organised pattern of manifestation). Phenomenal presentation – experiential character – arises only when mediational patterns achieve reflexive integration. The framework is relational, not panpsychist.

VI. Language

A29. Language as Finite-Bandwidth Synchronisation Protocol. Language is a coordination mechanism that compresses encounter-patterns into transmissible form and reconstructs approximate stabilisations in the receiving system. Communication is approximate synchronisation under constraint, not duplication.

A30. Convergence as Achievement, Not Default. Convergence is not the baseline condition of communication. It is a structurally constrained achievement that succeeds only when constraint-patterns are stable, mediational overlap is sufficient, and the domain admits of discrete restabilisation.

A31. The Ontological Grammar Thesis. The grammatical structures of natural language impose ontological commitments on everything they encode. Indo-European subject-predicate grammar scaffolds substance ontology by making object-first thinking fluent and relational or durative thinking cumbersome. Grammar does not imprison thought. It taxes alternatives.

A32. The Language Insufficiency Hypothesis. For any finite-bandwidth linguistic protocol operating across non-identical mediational architectures, there exist domains in which the constraint-patterns of encounter resist stable compression with sufficient fidelity to achieve reliable synchronisation. This is a structural feature of the protocol's operating conditions, not a defect of particular languages or speakers.

A33. Ontological Grammar Contributes to Bandwidth Degradation. In the Contestables and Fluids regions of the effectiveness gradient, language's ontological grammar actively mis-shapes what it encodes, producing systematic false convergence. The biases of the protocol contribute to its limits.

VII. Social Ontology

A34. Institutions as Second-Order Constraint Systems. Social constructs are second-order constraint systems sustained by coordinated behaviour, enforcement mechanisms, and material embedding. They alter real interactional trajectories regardless of participants' metaphysical beliefs. They are real because they constrain encounter.

A35. Institutional Reality Grounded in Constraint, Not Recognition. Institutions are not sustained by collective intentionality (Searle) but by coordinated behaviour that alters interactional trajectories. The constitutive mechanism depends on constraint, not on mental states.

A36. Institutional False Convergence. Institutional language routinely produces apparent coordination where actual synchronisation is absent. The protocol circulates compressed stabilisations that are decompressed differently by different agents, generating the appearance of agreement without its substance.

A37. Synchronisation Repair. Institutions persist not because their categories remain stable but because they continuously repair synchronisation failures through precedent, authoritative reinterpretation, and strategic ambiguity.

VIII. Realism

A38. Convergence Realism. MEOW's realism is grounded in the convergent stability of relational constraint across distinct mediational architectures. Invariance is inferred from convergence, not postulated as a hidden substrate. The framework is neither naïve object realism, nor constructivism, nor Kantian noumenalism, nor relativism.

A39. Graduated Realism. Realist warrant scales with convergent stability. Physics warrants strong realism. Biology warrants moderate realism. Social theory warrants cautious realism. Phenomenology warrants the most cautious realism of all. This graduation is not relativism. It is realism calibrated to evidence.

A40. Convergence Requires Genuine Heterogeneity. Strong convergence requires genuine diversity in mediational architecture and cross-context success in intervention. Convergence manufactured by shared protocol constraints (identical instruments, identical training) does not constitute evidence for invariance.

IX. Causation and Temporality

A41. Causation as Constraint Transmission. Causation is the stable sequencing of constraint-patterns across encounter-events. An encounter E_1 is causally relevant to E_2 when the constraint-profile of E_1 reliably restricts the possible configurations of E_2 across mediational variation.

A42. Temporal Organisation as Internal to Encounter. Temporal structure emerges within the organisation of mediation. Relations of simultaneity, succession, and duration are features of how encounters are patterned, not properties imposed from outside encounter.

X. Normativity (Open Frontier)

A43. The Normativity Gap. The framework describes what constraints obtain and what affordance gradients are available. It does not, by design, explain when constraint becomes obligation or when coordination becomes justice. Normativity is the principal open frontier.

A44. Necessary but Insufficient Conditions. Organismal persistence generates interests (but interests are too thin). Coordination pressure generates regularity (but regularity is too instrumental). Institutional enforcement generates compliance (but compliance is too circular). Each is necessary but not sufficient for normativity.

A45. The Framework Makes Normative Discourse Auditable. MEOW can diagnose false normativity, explain persistent moral disagreement as structurally predicted, and separate stable constraint-patterns from the moral narratives wrapped around them. It cannot generate prescriptions. It can inspect them.

Note on structural dependence. These commitments are mutually supporting rather than strictly hierarchical. However, certain commitments are structurally load-bearing: rejection of A1–A5 (the ontological primitives), A22 (knowledge as stabilisation), or A38 (convergence realism) would require substantial revision of the framework as a whole. The remaining commitments are derived from or dependent upon these central claims, and their revision would have local rather than global consequences.

Bibliography

Anscombe, G. E. M. (1975). *The first person*. In S. Guttenplan (Ed.), Mind and language (pp. 45–65). Clarendon Press.

Austin, J. L. (1962). *How to do things with words*. Oxford University Press.

Ayer, A. J. (1936). *Language, truth and logic*. Victor Gollancz.

Barad, K. (2007). *Meeting the universe halfway: Quantum physics and the entanglement of matter and meaning*. Duke University Press.

Beall, J. C., & Restall, G. (2006). *Logical pluralism*. Clarendon Press.

Bergson, H. (1910). *Time and free will: An essay on the immediate data of consciousness* (F. L. Pogson, Trans.). George Allen and Unwin. (Original work published 1889)

Bergson, H. (1911). *Creative evolution* (A. Mitchell, Trans.). Henry Holt and Company. (Original work published 1907)

Bergson, H. (1991). *Matter and memory* (N. M. Paul & W. S. Palmer, Trans.). Zone Books. (Original work published 1896)

Bhaskar, R. (2008). *A realist theory of science*. Routledge. (Original work published 1975)

Birkhoff, G., & von Neumann, J. (1936). The logic of quantum mechanics. *Annals of Mathematics*, 37(4), 823–843.

Blackburn, S. (1993). *Essays in quasi-realism*. Oxford University Press.

Blackburn, S. (1998). *Ruling passions: A theory of practical reasoning*. Clarendon Press.

Block, N. (1978). Troubles with functionalism. In C. W. Savage (Ed.), *Perception and cognition: Issues in the foundations of psychology* (pp. 261–325). University of Minnesota Press.

Bonhoeffer, D. (1971). *Letters and papers from prison* (Enlarged ed.). SCM Press. (Original work published 1953)

Boroditsky, L. (2001). Does language shape thought? Mandarin and English speakers' conceptions of time. *Cognitive Psychology*, 43(1), 1–22.

Boyd, R. N. (1983). On the current status of the issue of scientific realism. *Erkenntnis*, 19(1–3), 45–90.

Boyd, R. N. (1988). How to be a moral realist. In G. Sayre-McCord (Ed.), Essays on moral realism (pp. 181–228). Cornell University Press.

Brandom, R. B. (1994). *Making it explicit: Reasoning, representing, and discursive commitment*. Harvard University Press.

Brandom, R. B. (2000). *Articulating reasons: An introduction to inferentialism*. Harvard University Press.

Chakravartty, A. (2007). *A metaphysics for scientific realism: Knowing the unobservable*. Cambridge University Press.

Chalmers, D. J. (1996). *The conscious mind: In search of a fundamental theory*. Oxford University Press.

Chemero, A. (2003). An outline of a theory of affordances. *Ecological Psychology*, 15(2), 181–195.

Chemero, A. (2009). *Radical embodied cognitive science*. MIT Press.

Clark, A. (2013). Whatever next? Predictive brains, situated agents, and the future of cognitive science. *Behavioral and Brain Sciences*, 36(3), 181–204.

Clark, A. (2016). *Surfing uncertainty: Prediction, action, and the embodied mind*. Oxford University Press.

Collins, H. M. (1985). Changing order: Replication and induction in scientific practice. *Sage*.

Connolly, W. E. (1993). *The terms of political discourse* (3rd ed.). Princeton University Press.

Dennett, D. C. (1991). *Consciousness explained*. Little, Brown and Company.

Dennett, D. C. (1984). *Elbow room: The varieties of free will worth wanting*. MIT Press.

Dennett, D. C. (2003). *Freedom evolves*. Viking.

Dewey, J. (1938). *Logic: The theory of inquiry*. Henry Holt and Company.

Duhem, P. (1954). *The aim and structure of physical theory* (P. P. Wiener, Trans.). Princeton University Press. (Original work published 1906)

Dworkin, R. (1986). Law's empire. Harvard University Press.

Epstein, B. (2015). *The ant trap: Rebuilding the foundations of the social sciences*. Oxford University Press.

Esfeld, M. (2004). Quantum entanglement and a metaphysics of relations. *Studies in History and Philosophy of Modern Physics*, 35(4), 601–617.

Evans, N., & Levinson, S. C. (2009). The myth of language universals: Language diversity and its importance for cognitive science. *Behavioral and Brain Sciences*, 32(5), 429–448.

Field, H. (2001). *Truth and the absence of fact*. Clarendon Press.

Foucault, M. (1977). *Discipline and punish: The birth of the prison* (A. Sheridan, Trans.). Pantheon Books. (Original work published 1975)

Frankish, K. (2016). Illusionism as a theory of consciousness. *Journal of Consciousness Studies*, 23(11–12), 11–39.

French, S. (2014). *The structure of the world: Metaphysics and representation*. Oxford University Press.

Friston, K. (2010). The free-energy principle: A unified brain theory? *Nature Reviews Neuroscience*, 11(2), 127–138.

Garfield, J. L. (1995). *The fundamental wisdom of the middle way: Nāgārjuna's Mūlamadhyamakakārikā*. Oxford University Press.

Gallie, W. B. (1956). Essentially contested concepts. *Proceedings of the Aristotelian Society*, 56, 167–198.

Geach, P. T. (1960). Ascriptivism. *The Philosophical Review*, 69(2), 221–225.

Geach, P. T. (1965). Assertion. *The Philosophical Review*, 74(4), 449–465.

Gibbard, A. (1990). *Wise choices, apt feelings: A theory of normative judgment*. Harvard University Press.

Gibson, J. J. (1979). *The ecological approach to visual perception*. Houghton Mifflin.

Grice, H. P. (1989). *Studies in the way of words*. Harvard University Press.

Habermas, J. (1984). *The theory of communicative action: Volume 1. Reason and the rationalization of society* (T. McCarthy, Trans.). Beacon Press.

Hacking, I. (1983). *Representing and intervening: Introductory topics in the philosophy of natural science*. Cambridge University Press.

Hanson, N. R. (1958). *Patterns of discovery: An inquiry into the conceptual foundations of science*. Cambridge University Press.

Hare, R. M. (1952). *The language of morals*. Clarendon Press.

Hart, H. L. A. (1961). *The concept of law*. Clarendon Press.

Heidegger, M. (1962). *Being and time* (J. Macquarrie & E. Robinson, Trans.). Blackwell. (Original work published 1927)

Horwich, P. (1990). *Truth*. Clarendon Press.

Hohwy, J. (2013). *The predictive mind*. Oxford University Press.

Hume, D. (2000). *A treatise of human nature* (D. F. Norton & M. J. Norton, Eds.). Oxford University Press. (Original work published 1739)

Husserl, E. (1982). *Ideas pertaining to a pure phenomenology and to a phenomenological philosophy*, first book (F. Kersten, Trans.). Martinus Nijhoff. (Original work published 1913)

Hutto, D. D., & Myin, E. (2013). *Radicalizing enactivism: Basic minds without content*. MIT Press.

James, W. (1975). *Pragmatism: A new name for some old ways of thinking*. Harvard University Press. (Original work published 1907)

Jonas, H. (1966). *The phenomenon of life: Toward a philosophical biology*. Harper & Row.

Kant, I. (1998). *Critique of pure reason* (P. Guyer & A. W. Wood, Trans.). Cambridge University Press. (Original work published 1781)

Korzybski, A. (1933). Science and sanity: An introduction to non-Aristotelian systems and general semantics. *International Non-Aristotelian Library*.

Kripke, S. A. (1980). *Naming and necessity*. Harvard University Press.

Köhler, W. (1947). Gestalt psychology: An introduction to new concepts in modern psychology. Liveright.

Kuhn, T. S. (2012). *The structure of scientific revolutions* (4th ed.). University of Chicago Press. (Original work published 1962)

Ladyman, J., & Ross, D. (2007). *Every thing must go: Metaphysics naturalized*. Oxford University Press.

Latour, B. (2005). *Reassembling the social: An introduction to actor-network-theory*. Oxford University Press.

Levine, J. (1983). Materialism and qualia: The explanatory gap. *Pacific Philosophical Quarterly*, 64(4), 354–361.

Lewin, K. (1951). *Field theory in social science: Selected theoretical papers*. Harper & Brothers.

Cited in: Ch2 FN1 (Gestalt field theory)

Locke, J. (1975). *An essay concerning human understanding* (P. H. Nidditch, Ed.). Clarendon Press. (Original work published 1690)

Lucy, J. A. (1992). *Grammatical categories and cognition: A case study of the linguistic relativity hypothesis*. Cambridge University Press.

Matilal, B. K. (1986). *Perception: An essay on classical Indian theories of knowledge*. Oxford University Press.

Maturana, H. R., & Varela, F. J. (1980). *Autopoiesis and cognition: The realization of the living*. D. Reidel.

Merleau-Ponty, M. (1962). *Phenomenology of perception* (C. Smith, Trans.). Routledge. (Original work published 1945)

Mumford, S., & Anjum, R. L. (2011). *Getting causes from powers.* Oxford University Press.

Nagel, T. (1986). *The view from nowhere.* Oxford University Press.

Nagel, T. (1974). What is it like to be a bat? *The Philosophical Review,* 83(4), 435–450.

Noë, A. (2004). *Action in perception.* MIT Press.

Parfit, D. (1984). *Reasons and persons.* Clarendon Press.

Peirce, C. S. (1878). How to make our ideas clear. *Popular Science Monthly,* 12, 286–302.

Perry, J. (1979). The problem of the essential indexical. *Noûs,* 13(1), 3–21.

Pinker, S. (2007). *The stuff of thought: Language as a window into human nature.* Viking.

Priest, G. (2006). *In contradiction: A study of the transconsistent* (2nd ed.). Oxford University Press.

Putnam, H. (1968). Is logic empirical? In R. S. Cohen & M. W. Wartofsky (Eds.), *Boston studies in the philosophy of science* (Vol. 5, pp. 216–241). D. Reidel.

Putnam, H. (1975). The meaning of 'meaning'. In K. Gunderson (Ed.), *Language, mind, and knowledge* (pp. 131–193). University of Minnesota Press.

Putnam, H. (1981). *Reason, truth, and history.* Cambridge University Press.

Quine, W. V. O. (1969). *Ontological relativity and other essays.* Columbia University Press.

Quine, W. V. O. (1951). Two dogmas of empiricism. *The Philosophical Review,* 60(1), 20–43.

Reed, E. S. (1996). Encountering the world: Toward an ecological psychology. Oxford University Press.

Rescher, N. (1996). *Process metaphysics: An introduction to process philosophy.* SUNY Press.

Rorty, R. (1979). *Philosophy and the mirror of nature.* Princeton University Press.

Roskies, A. L. (1999). The binding problem. *Neuron*, 24(1), 7–9.

Rovelli, C. (1996). Relational quantum mechanics. *International Journal of Theoretical Physics,* 35(8), 1637–1678.

Sapir, E. (1929). The status of linguistics as a science. *Language*, 5(4), 207–214.

Schwitzgebel, E. (2016). Phenomenal consciousness, defined and defended as innocently as I can manage. *Journal of Consciousness Studies*, 23(11–12), 224–235.

Scott, J. C. (1998). *Seeing like a state: How certain schemes to improve the human condition have failed*. Yale University Press.

Searle, J. R. (1992). *The rediscovery of the mind*. MIT Press.

Searle, J. R. (1995). *The construction of social reality*. Free Press.

Searle, J. R. (2010). *Making the social world: The structure of human civilization*. Oxford University Press.

Sellars, W. (1963). *Science, perception and reality*. Ridgeview Publishing.

Sellars, W. (1997). *Empiricism and the philosophy of mind*. Harvard University Press. (Original work published 1956)

Shannon, C. E. (1948). *A mathematical theory of communication*. Bell System Technical Journal, 27(3), 379–423.

Singer, W., & Gray, C. M. (1995). Visual feature integration and the temporal correlation hypothesis. *Annual Review of Neuroscience*, 18, 555–586.

Sperber, D., & Wilson, D. (1986). *Relevance: Communication and cognition.* Harvard University Press.

Stanford, P. K. (2006). *Exceeding our grasp: Science, history, and the problem of unconceived alternatives*. Oxford University Press.

Stevenson, C. L. (1944). *Ethics and language*. Yale University Press.

Strawson, G. (2006). Realistic monism: Why physicalism entails panpsychism. *Journal of Consciousness Studies*, 13(10–11), 3–31.

Sturgeon, N. L. (1985). Moral explanations. In D. Copp & D. Zimmerman (Eds.), *Morality, reason and truth* (pp. 49–78). Rowman & Allanheld.

Thompson, E. (2007). *Mind in life: Biology, phenomenology, and the sciences of mind*. Harvard University Press.

Treisman, A. (1996). The binding problem. Current Opinion in Neurobiology, 6(2), 171–178.

Turing, A. M. (1950). Computing machinery and intelligence. *Mind*, 59(236), 433–460.

Turner, D. (1995). *The darkness of God: Negativity in Christian mysticism*. Cambridge University Press.

Turvey, M. T. (1992). Affordances and prospective control: An outline of the ontology. *Ecological Psychology*, 4(3), 173–187.

Uexküll, J. von. (2010). *A foray into the worlds of animals and humans, with A theory of meaning* (J. D. O'Neil, Trans.). University of Minnesota Press. (Original work published 1934)

van Fraassen, B. C. (1980). *The scientific image*. Oxford University Press.

Varela, F. J., Thompson, E., & Rosch, E. (1991). *The embodied mind: Cognitive science and human experience*. MIT Press.

Whitehead, A. N. (1978). *Process and reality: An essay in cosmology* (Corrected ed.). Free Press. (Original work published 1929)

Whorf, B. L. (1956). *Language, thought, and reality: Selected writings*. MIT Press.

Willis, B. (2025). The mediated encounter ontology of the world: A relational metaphysics beyond mind and world. https://doi.org/10.5281/zenodo.17685689

Willis, B. (2026). *A language insufficiency hypothesis: Mapping the boundaries of linguistic expression*. Philosophics Press.

Willis, B. (2026). *When language fails: Ontological grammars and the limits of articulation*. Philosophics Press.

Wittgenstein, L. (2009). *Philosophical investigations* (4th ed.). Wiley-Blackwell. (Original work published 1953)

Zipf, G. K. (1949). *Human behavior and the principle of least effort*. Addison-Wesley.

NB: The following essay is a facsimile of the oringinal MEOW essay, It is included here as a convenience. Original page formatting has been retained at the expense of page count. A canonical copy of *The Mediated Encounter Ontology of the World* is available on Zenado as a PDF. https://doi.org/10.5281/zenodo.17685689

The Mediated Encounter Ontology of the World:
A Relational Metaphysics Beyond Mind and World

Bry Willis

Independent Scholar

Author Information and Disclosure

ORCID iD: https://orcid.org/0009-0005-0016-5694

This essay represents independent research; the author has no institutional affiliation.

The author declares no conflict of interest.

This research received no specific grant from any funding agency in the public, commercial, or not-for-profit sectors.

Correspondence concerning this article should be addressed to microglyphics@gmail.com.

Permanent identifier: https://doi.org/10.5281/zenodo.17685689

Abstract

This paper argues that the realism–idealism debate rests on a false dichotomy sustained by inherited substance ontology, and that the debate dissolves once encounter-events are taken as ontologically primary. The realism–idealism debate has reached a productive impasse: realism cannot explain how fully mediated access could ever disclose a mind-independent reality, while idealism struggles to account for the genuine resistance encountered in experience. This paper argues that both positions falter because they inherit a shared architectural assumption—that "mind" and "world" are independently existing substances that subsequently enter into relation. I develop the Mediated Encounter Ontology of the World (MEOW) as an alternative framework in which encounter-events are ontologically basic. An encounter-event is a structured relational occurrence in which mediation, presentational structure, and constraint co-emerge; "mind" and "world" are derivative abstractions extracted from stable patterns across such events.

The argument unfolds in four movements: diagnostic (the failure of substance ontology), constructive (the architecture of encounter-events), genealogical (situating MEOW within relational traditions), and applicative (consequences for epistemology, science, and cognition). First, I show that perceptual, cognitive, and linguistic–cultural mediation is not an epistemic obstacle but the constitutive condition under which any encounter becomes possible. Second, I analyse the intrinsic forms of constraint that render encounters neither arbitrary nor freely malleable, thereby avoiding any appeal to a noumenal substrate. Third, I locate MEOW within a broader relational lineage—phenomenology, pragmatism, enactivism, and nondual traditions—while distinguishing it from adjacent positions such as neutral monism, panpsychism, and process metaphysics. Finally, I outline the principal applicative implications of the framework, showing how MEOW reconfigures debates in epistemology, scientific explanation, consciousness, and extended cognition without relying on a standpoint outside mediation.

MEOW dissolves the realism–idealism dichotomy without collapsing into relativism or subjectivism. By taking relational encounter as ontologically primary, it preserves empirical constraint without invoking unmediated access and offers new conceptual resources for epistemology, philosophy of science, cognitive theory, and the metaphysics of consciousness.

Keywords: relational ontology; encounter-events; mediation; constraint; realism–idealism debate; anti-substantivalism; phenomenology; enactivism; extended cognition; philosophy of science; consciousness; perspectival realism

I. Introduction

The long-standing dispute between realism and idealism has reached a stage of diminishing returns. Realism insists that thought and experience answer to a mind-independent world, yet it cannot explain how such a world is accessed without presupposing the very unmediated vantage it must deny. Idealism reverses the polarity, grounding reality in mind, experience, or conceptual structure, yet it struggles to account for the persistent forms of resistance—physical, cognitive, conceptual—that frustrate subjective projection. Both traditions inherit a deeper assumption: that "mind" and "world" exist as separable domains that subsequently enter into epistemic relation. This presupposition reproduces a familiar architecture of false contrasts—appearance versus reality, representation versus construction, inner versus outer—that continues to structure contemporary debates long after its usefulness has waned.

This paper develops the Mediated Encounter Ontology of the World (MEOW) as an alternative to that inherited architecture. MEOW rejects the idea of pre-existing substances waiting to be bridged. It begins instead with *encounter-events*: structured relational occurrences in which mediation, presentational structure, and constraint arise together. On this view, there is no metaphysical gap to be crossed by perception or representation; there is only the event of encounter itself, already mediated, already meaningful, already resistive. "Mind" and "world" are not ontological primitives but abstractions—patterns distilled from regularities across sequences of encounter-events.

The motivation for this reorientation is straightforward. Our access to reality is triply mediated: biologically, through species-specific sensory and motor systems; cognitively, through predictive and inferential structure; and linguistically–culturally, through inherited schemes of classification and interpretation. Mediation is not an epistemic blemish but a condition of possibility for any encounter. At the same time, experience exhibits stable forms of constraint—

physical resistance, cognitive recalcitrance, cross-perspectival invariance—that cannot be dissolved into projection without eliminating the very distinction between success and error. Any viable metaphysics must therefore accommodate both pervasive mediation and genuine constraint without appealing to unobservable substrates posited behind experience.

The claim advanced here is that such accommodation is achieved not by refining realism or rehabilitating idealism, but by discarding their shared assumption. If encounter-events are ontologically basic, there is no need for substances standing behind them. The explanatory burdens once borne by "mind" and "world" are redistributed across the architecture of mediation and constraint internal to encounter itself. MEOW thus dissolves the realism–idealism dichotomy rather than adjudicating it, offering a unified framework in which epistemic, scientific, and phenomenological practices can be understood without invoking a standpoint outside mediation.

The remainder of the paper proceeds as follows. Section II examines the motivating problems: the structure of mediation, the varieties of experiential resistance, and the failure of substance-based ontologies to accommodate their coexistence. Section III articulates the core commitments of MEOW, defining encounter-events and analysing their internal organisation. Section IV situates MEOW within a broader relational genealogy while distinguishing it from adjacent positions. Section V clarifies what MEOW is not, and Section VI outlines its implications for epistemology, philosophy of science, cognition, and consciousness. Section VII concludes by summarising the reorientation proposed here and indicating directions for further development.

II. Genesis of the Framework

The contemporary impasse between realism and idealism is not a misunderstanding, nor an honest disagreement between two viable options. It is the predictable by-product of a shared conceptual inheritance. Both traditions presuppose a subject–object ontology in which "mind" and "world" are treated as independently existing substances that later enter into epistemic relation. Realism maintains that a mind-independent world grounds objectivity; idealism maintains that the structuring activity of mind is primary. For all their opposition, both accept the same architecture: cognition is a relation between two already constituted domains. It is this inherited bifurcation that generates the familiar tensions neither camp can resolve without undermining its own foundations.

Two features of experience expose the inadequacy of this architecture. The first is pervasive mediation. At every level, epistemic access is structurally organised—through species-specific perceptual systems, predictive and inferential cognitive dynamics, and linguistically mediated conceptual schemas. Experience is never encountered "as it is"; it is always shaped, patterned, and filtered through biological, cognitive, and cultural organisation. Any metaphysics that demands unmediated access—whether as an ideal limit or a regulative postulate—faces an impossible choice: either weaken its claim to mind-independence or appeal to an inaccessible substrate that explains nothing. Realism thus oscillates between modesty (surrendering its original promise) and dogmatism (invoking worlds that cannot, in principle, be known).

The second is genuine resistance. However thoroughly organised experience may be, it is not infinitely plastic. Certain patterns persist regardless of framing: gravitational effects that ignore conceptual preference; toxic substances that remain toxic despite linguistic reframing; perceptual limitations that cannot be trained or willed away; cross-perspectival invariances systematically uncovered by scientific practice. Resistance is not an artefact of projection nor an interpretive convenience. It is a structural feature of experience that idealism cannot dissolve without forfeiting explanatory scope or collapsing into the realism it set out to reject.

Individually, mediation challenges realism; resistance challenges idealism. Taken together, they reveal the deeper problem: both traditions inherit a dichotomy—mind here, world there—that lived experience does not endorse. Mediation becomes a threat to the realism that requires unmediated contact; resistance becomes a threat to the idealism that requires constructive sovereignty. The resulting oscillation between these poles is not evidence of insufficient ingenuity but of an inherited conceptual frame that forces an impossible choice.

The need for an alternative emerges not from dissatisfaction with either tradition's achievements but from recognition that the categories themselves are misdrawn. A viable metaphysics must integrate mediation and resistance without presupposing the very split that renders them incompatible. The genesis of the Mediated Encounter Ontology of the World (MEOW) lies in this diagnostic insight: the realism–idealism debate persists because it is framed by assumptions that experience does not, in fact, support.

The conclusion is therefore straightforward. If mediation and resistance are basic features of our engagement with reality, then the primary unit of analysis cannot be independently existing substances but encounter-events—structured relational occurrences in which these features arise together. The next section develops a metaphysical framework that begins where the inherited dichotomy does not: with encounter rather than substance, with relation rather than relata.

III. Articulating the Mediated Encounter Ontology of the World

III.A. Encounter-Events: Ontological Priority

The Mediated Encounter Ontology of the World (MEOW) begins from a simple but decisive reorientation: the most fundamental units of reality are not substances, objects, subjects, or mental states, but encounter-events. An encounter-event is a structured relational occurrence in which mediation, experiential presentation, and constraint arise together. It is not a relation between antecedently existing relata. Rather, relata are abstractions retrospectively distilled from stable patterns across many such events. To see a tree, grasp a concept, or measure a physical system is not to connect two independently existing domains, but to participate in a single organised event of encounter.

This reversal of priority shifts the explanatory burden. In substance-based metaphysics, relations derive from relata; in MEOW, relata derive from patterns within networks of encounter. The familiar categories of "mental" and "physical" are therefore not ontological kinds but aspectual designations applied to differing modes of encounter-organisation. Encounter-events are not mental episodes awaiting external interpretation, nor physical processes awaiting internal representation; they are the elementary structure within which mediation, presentation, and constraint co-emerge.

III.A.1. Individuation of Encounter-Events

Encounter-events are individuated not by substances or enduring objects, but by the organisation of mediation, presentational structure, and constraint that constitutes them. Two encounter-events are numerically distinct when their mediational–resistive patterns differ in structure, stability, or functional role. This criterion is structural rather than substantial: it concerns how an event is organised, not what it is "made of."

Because mediational architectures are layered—biological, cognitive, cultural—encounter-events can be nested, overlapping, or extended, depending on which organisational patterns are being tracked. A single perceptual scene may contain sub-events (edge-detection, colour segmentation, object-binding), while multiple observers can participate in distinct but structurally related encounters that share a constraint profile. Temporal boundaries are likewise pragmatic: an encounter-event may span milliseconds (synaptic integration), seconds (visual fixation), or hours (scientific measurement) insofar as a coherent mediational–resistive organisation persists across that duration.

This individuation scheme is intentionally pluralistic. MEOW does not impose a single metaphysical criterion for carving experience into events, because the demand for such a criterion presupposes substance ontology. Instead, encounter-boundaries are fixed by the pragmatic coherence of the mediational and constraining structures relevant to explanation. An encounter may be treated as unified for one purpose (e.g., "measuring this sample") and decomposed for another (e.g., analysing error sources or perceptual steps). This is not relativism but recognition that relational organisation—not absolute metaphysical segmentation—determines what counts as an event.

Importantly, encounter-events do not require consciousness. Presentational structure—the organised pattern of how an encounter is manifested—exists even in non-conscious encounters such as chemical reactions, tectonic shifts, or gravitational interactions. Phenomenal presentation, by contrast, is a specialised form of presentational structure that arises only when mediational patterns are reflexively integrated (as in biological cognition). This distinction preserves MEOW's commitment to non-substantival relational ontology without sliding into panpsychism: all encounters exhibit organisation, but only some exhibit experience.

On this view, individuation is neither arbitrary nor metaphysically rigid. It is grounded in the stability, coherence, and functional role of mediational–resistive organisation. Encounter-events are thus the basic units of MEOW not because they are smallest or simplest, but because they are the loci at which mediation and constraint cohere into structured reality.

III.A.2 Non-Conscious Encounter-Structure

MEOW does not posit subjectless experience. Experience, in the phenomenal sense, arises only within mediational architectures capable of reflexive integration (biological or otherwise). What MEOW does require is subjectless encounter-structure: the organised pattern of mediation and constraint that *could* manifest within an appropriate mediational system, even if none is currently present.

Two clarifications follow.

First, encounter-events do not imply two pre-existing relata "meeting." "Encounter" names the relational organisation through which relata become identifiable as patterns. Chemical reactions, tectonic drift, and photon–detector interactions exhibit encounter-structure because they instantiate stable mediational–resistive organisation, not because they resemble interpersonal meetings.

Second, non-conscious encounters possess presentational structure, not phenomenal presentation. Presentational structure is the way an encounter is organised such that it is *manifestable* to a suitable mediational system. A tectonic shift has organised strain patterns that would, if coupled to an appropriate sensory or computational architecture, appear as measurable displacement or pressure gradients. These patterns do not require a subject to exist; they require one to be phenomenally disclosed.

Thus encounter-structure is broader than experience without collapsing into panpsychism or reductive physicalism. MEOW treats relational organisation—not consciousness and not substance—as ontologically primary. Experience is one mode of that organisation; it is not its measure.

III.B. Mediation as Ontological Condition

Within MEOW, mediation is not an epistemic impediment but the structural condition under which any encounter-event arises. Perceptual interfaces, sensorimotor capacities, predictive cognitive architectures, and the linguistic–conceptual frameworks inherited from cultural participation jointly determine how an encounter is organised. These mediational patterns are not filters that stand between an experiencer and a world; they are the very organisation through which reality is disclosed.

Because mediation is intrinsic to the structure of encounter-events, it cannot be bracketed, bypassed, or purified. There is no "raw" access to be reclaimed, no metaphysical vantage untainted by biological or conceptual organisation. Mediation is not a distortion of reality but its mode of manifestation: the way reality comes to presence in encounter. This applies equally to conscious and non-conscious encounters. In both cases the mediational substrate—whether neural, chemical, mechanical, or computational—determines the organisation of the event's *presentational structure*, even though only some encounters give rise to *phenomenal presentation*.

The philosophical significance of this point is straightforward. If every encounter is structurally mediated, then the aspiration to reach an unmediated reality behind experience is not merely impossible in practice—it is conceptually incoherent. Unmediated access could only be access without organisation, which would be indistinguishable from no access at all. Mediation is the condition that makes encounter possible, not a veil to be stripped away.

This reorientation defuses a long-standing tension in metaphysics. Realism requires unmediated contact to ground mind-independent objectivity, while idealism risks reducing constraint to the activity of mind or concept. MEOW avoids this dilemma by treating mediation as ontological rather than epistemic: encounter-events are constituted through structured mediation, and this mediation is part of what reality *is*, not an obstacle to knowing it. The world does not stand behind mediation; the world appears through mediation.

III.C. Resistance as Internal Constraint

Where classical idealisms often interpret resistance as an inconvenience to be explained away, and realisms regard it as evidence for an external world never directly encountered, MEOW locates resistance within the organisation of the encounter-event itself. Constraint—what earlier drafts called "resistance"—is the dimension of an encounter that is not freely variable. It is the patterned invariance, pushback, or recalcitrance that emerges whenever a mediational structure meets a structured field of possibilities.

Constraint is not imported from a noumenal realm nor projected by a subjective mind. It arises from the interplay between the mediational architecture of an encounter and the structural regularities that persist across such encounters. The grain of wood under a plane, the gravitational pull on a falling body, the persistence of the Müller–Lyer illusion, the impossibility of conceptual contradictions—each is a case in which the encounter exhibits a stable pattern that cannot be willed, reframed, or interpreted out of existence. Constraint is thus internal to the event's organisation: it is how the world's patterned stability is manifested *through* mediation rather than behind it.

This framing allows MEOW to preserve empirical constraint without invoking unmediated access to a mind-independent substrate. Constraint does not signal an external object exerting influence from outside the encounter-event; it signals the structure that makes divergent encounters converge, that grounds correction and error, and that enables cross-perspectival coherence. Mediation shapes how an encounter is organised, but constraint ensures that encounters are not arbitrary.

In this respect, MEOW replaces the realist's hidden noumenon and the idealist's pliable construct with a relational ontology in which constraint is intrinsic and indispensable. The experience of pushback—physical, cognitive, conceptual—is not evidence for something "out there" nor merely something "in here." It is evidence that encounter-events have an organised structure in which not everything is up for negotiation.

With mediation and constraint understood as co-constitutive, MEOW eliminates the need to choose between independence and construction. What appears as resistance is simply the encounter's structured stability—its refusal to collapse into projection, and its refusal to resolve into inaccessible substrate. Constraint is thus neither brute externality nor sovereign construction; it is the internal regularity that makes encounter-events intelligible, corrigible, and shared.

III.D. The Internal Organisation of Encounter-Events

Encounter-events are structured through the co-emergence of mediation, presentational structure, and constraint. To understand how these aspects function together—and why neither realism nor idealism can accommodate them—a more precise account is required. This section outlines the internal logic of encounter-organisation before turning to illustrative cases.

III.D.1 Structural Logic: Mediation, Presentational Structure, Constraint

The internal architecture of an encounter-event has three inseparable aspects:

1. Mediation determines the *modal shape* of an encounter—what kinds of structure can be registered, how information is organised, and how possibilities are discriminated. It comprises biological, cognitive, and cultural organisation.
2. Constraint determines the *stability* of an encounter—the patterns that persist across variation in perspective, framing, or expectation. Constraint appears as resistance, invariance, or recalcitrance.
3. Presentational structure is the way the encounter is *manifested*. In conscious encounters, this may include phenomenal presentation ("what it is like"). In non-conscious encounters, it consists of the organised pattern through which mediation and constraint are instantiated, without any corresponding phenomenal character.

These three aspects do not cause one another nor stand in temporal sequence. They are structurally entailed within a single relational event. Mediation always structures some field of possibility; structured fields always display constraint; and constraint can only appear within a mediational pattern. Presentational structure is the unified manifestation of that organisation.

This co-emergence is not mysterious. It simply means that the relata presupposed by substance metaphysics—"mind" and "world"—are not prior components but aspectual descriptions of the event's internal organisation. Different descriptions emphasise different aspects of the same structure. MEOW's central claim is that nothing metaphysically deeper is needed.

III.D.2 Worked Example: Colour Encounter

Colour perception demonstrates the internal organisation of encounter-events with unusual clarity.

Mediation:

Several layers jointly determine the structure of a colour-encounter:

- Biological mediation: the trichromatic cone system, opponent processes, retinal adaptation.
- Cognitive mediation: predictive models, memory colour, categorical expectations.
- Cultural–linguistic mediation: colour vocabularies and learned category boundaries shape categorical expectations (Berlin & Kay, 1969; Roberson, Davies, & Davidoff, 2000).

These jointly shape how colour appears—not merely "how it is interpreted," but how the world manifests in the encounter.

Constraint:

Despite this variability, certain patterns are not freely alterable:

- Metameric matches remain stable even when expectations change.
- Focal colours (e.g., best examples of red or blue) display cross-cultural robustness (Berlin & Kay, 1969).
- Cross-cultural variation in colour categories does not eliminate shared perceptual invariances, including focal colour salience and discrimination patterns (Roberson, Davies, & Davidoff, 2000).
- The Müller–Lyer–type depth illusions in colour–lightness interactions persist despite conscious knowledge.
- Wavelength sensitivities remain invariant regardless of linguistic categories.

Constraint is not imported from a hidden external world nor generated by conceptual activity. It emerges from the organised relation between mediational structure and environmental regularities.

Presentational Structure:

A colour-encounter manifests a single integrated field of appearance—a continuously structured pattern of hue, saturation, brightness—regardless of whether the subject is aware of the mediation producing it. In conscious perception, this includes phenomenal character; in computational colour detection (machine vision), it includes only presentational structure.

Why This Matters for MEOW

Colour encounters show that:

- variation in mediation does *not* eliminate constraint
- constraint does *not* require unmediated access
- cross-perspectival stability is possible without noumenal realism

The phenomenon arises through the organisation of the encounter-event, not through the interaction of two independent substances.

III.D.3 Worked Example: Radiometric Dating and Scientific Encounter

Scientific encounters exhibit the same internal organisation at a more abstract scale. See Dalrymple (1991) for a standard overview of radiometric methods.

Mediation:

- Instrumentation: detector sensitivity, calibration curves, decay-model assumptions
- Data processing: Bayesian inference, statistical filtering
- Conceptual frameworks: geological timescales, nuclear physics, error theory

These determine how radiometric signals are detected, processed, and interpreted.

Constraint:

Radiometric decay rates exhibit stable cross-perspectival regularities:

- half-lives remain constant across instruments, conditions, and conceptual framings
- independent isotopic systems converge on the same age estimates
- discrepancies allow identification of contamination rather than undermining the method

Again, the stability is not a peek behind mediation but a pattern within the structure of mediated encounters.

Presentational Structure:

The encounter here is not visual but quantitative: a structured field of numerical readings, error bounds, confidence intervals. Human or machine interpretation is not added afterwards; it is part of the encounter's organisation.

Why This Matters for MEOW

Radiometric dating shows that:

- objectivity arises from convergence across differently mediated encounters
- scientific inference does not require access to a world "behind" experience
- past-tense claims (e.g., "dinosaurs existed 65 million years ago") depend on constraint-patterns in present encounters

This directly addresses the realism worry without capitulating to idealism.

III.D.4 Generalising from the Examples

These examples demonstrate the general MEOW principle:

Mediation determines the *shape* of an encounter; constraint determines its *stability*; presentational structure is their unified manifestation.

Because these three aspects co-emerge, neither realism nor idealism can claim them as foundational. They only appear as "opposed" when one presupposes a subject–object split. Within MEOW, they are differentiated aspects of a single structured event.

III.E. Emergent Abstractions: Mind and World

If encounter-events are ontologically primary, then "mind" and "world" are not pre-existing domains but abstractions distilled from patterns across many such events. They are not substances but stabilities—regularities that emerge when sequences of encounters display characteristic organisations.

1. The Abstraction of Mind

"Mind" is the name we give to recurrent mediational–presentational patterns that exhibit:

- reflexive integration (encounters incorporating their own prior states)
- inference, prediction, and memory
- coherence across temporally extended encounters
- intra-encounter differentiation (self–other, figure–ground, intention–action)

These patterns cluster around a certain class of biological encounter-systems, but nothing metaphysical hinges on their biological instantiation. "Mind" identifies a stable organisational profile, not a substance that stands behind the profile.

In this sense, mind is emergent: the result of patterned regularities across encounter-networks, not a primitive ontological kind.

2. The Abstraction of World

"World," analogously, names the constraint-dominated regularities that persist across variations in mediation:

- cross-perspectival invariances
- stable causal–temporal sequences
- patterns that resist variation across biological, cultural, or instrumental differences
- constraint structures exhibited in physical, chemical, and mathematical regularities

To call something "worldly" is not to posit a substance behind experience. It is to say:

This pattern of constraint recurs across encounter-events irrespective of the mediational differences between them.

This is the operative meaning of objectivity in MEOW: stability across mediation, not access to an unmediated substrate.

3. Past Entities and Temporal Projection

MEOW handles past-tense claims without invoking noumena. Statements such as "dinosaurs existed" are grounded in constraint-patterns of present encounter-events:

- fossil morphology
- decay curves

- geological stratification
- cross-confirming isotopic systems

These present constraints support *retrodictive inference*—not to mind-independent substances, but to temporal patterns consistent with the constraint profiles we currently observe. Nothing in MEOW requires dinosaurs to be "encountered as they were," only that their inferred patterns are anchored in present stability.

Temporal projection is thus a form of mediated reasoning over constraint-patterns, not a leap outside encounter.

4. *Distant and Inaccessible Phenomena*

Similar reasoning applies to regions beyond direct encounter:

- exoplanets
- early-universe epochs
- black hole interiors
- unobservable subatomic states

Claims about them are not metaphysical commitments to inaccessible substances but extrapolations from constraint-patterns within actual encounters (spectral lines, gravitational lensing, cosmic background anisotropies, perturbation models).

What is "real" in MEOW is not the unreachable substrate but the network of constraint-relations that support inference.

5. *Counterfactuals and Objectivity*

Counterfactual reasoning ("If the isotope had been contaminated...") also emerges naturally:

- counterfactuals = variations in mediational pathways
- grounded by = stable constraints that predictably limit those variations

Thus, objectivity is not correspondence to an external world but the stability of constraint-patterns across possible mediational reorganisations.

6. *Why This Matters*

This expanded treatment dissolves a persistent misunderstanding:

- MEOW *does* accommodate durable structures
- MEOW *does* explain the stability required for science
- MEOW *does* handle the temporal, spatial, and modal scope of ordinary and scientific realism
- but without positing world-substances beyond encounter or mind-substances behind mediation

"Mind" and "world" are useful abstractions—indispensable, even—but they track recurring profiles of encounter, not independently existing domains.

III.F. Formal Summary of MEOW's Commitments

The central commitments of the Mediated Encounter Ontology of the World can be stated in a compact set of theses. These are not additional claims but a condensation of the framework developed in the preceding sections.

Thesis 1. Ontological Primacy of Encounter-Events

All that exists exists through encounter-events (E): structured relational occurrences in which mediation, presentational structure, and constraint co-emerge. Relata are abstractions over patterns within such events, not ontological primitives.

Formal:

$\forall x$ (x exists $\rightarrow$ x participates in some E).

There are no independently existing substances whose relations are subsequently added.

Thesis 2. Mediation as Constitutive Structure

Every encounter-event is organised through mediation—biological, cognitive, and linguistic–cultural. Mediation is not a veil between mind and world but the structural condition under which encounter becomes possible.

Formal:

If E is an encounter-event, then E entails M (structured mediation).

No access occurs outside mediation; "unmediated contact" is incoherent.

Thesis 3. Constraint as Intrinsic Stability

Constraint arises within encounter-events themselves. It is the dimension of stability, resistance, or invariance that is not freely alterable by variation in mediation. Constraint is not projection and not a noumenal intrusion.

Formal:

If E is an encounter-event, then E entails C (constraint), where C ≠ projection and C ≠ unmediated substrate.

Thesis 4. Co-Emergence of Mediation, Presentational Structure, and Constraint

Mediation, presentational structure, and constraint are not sequential components but aspectual descriptions of a single relational organisation.

Formal:

For any encounter-event E:

$E = \langle M, P, C \rangle$,

where no aspect is metaphysically prior and all arise jointly in the event's organisation.

(P = presentational structure; phenomenal presentation is a special case of P.)

Thesis 5. Derivative Status of "Mind" and "World"

"Mind" and "world" designate stable patterns abstracted from regularities across encounter-events rather than independently existing substances.

Formal:

Mind = abstraction over recurrent encounter-patterns dominated by M and P.

World = abstraction over recurrent encounter-patterns dominated by C.

Neither term denotes a substance external to E.

Thesis 6. Rejection of the Mental–Physical Dichotomy

"Mental" and "physical" are aspectual predicates applied to different emphases within encounter-event organisation. They identify classificatory stances, not ontological kinds.

Formal:

∀x: mental(x) and physical(x) are aspectual descriptions of encounter-structure, not substance predicates.

Thesis 7. Relational Closure

All experiential, epistemic, and causal processes occur within networks of encounter-events. There is no need—and no conceptual room—for positing extra-encounter substances or forces.

Formal:

All processes supervene on networks of encounter-events;

no extra-encounter causation exists.

Thesis 8. Causal Dependence as Stable Encounter Sequencing

Causation is not substance-level production but the stability of constraint transmission across sequences of encounter-events. An encounter E_1 is causally relevant to E_2 when the constraint-profile of E_1 reliably shapes, limits, or organises the possible configurations of E_2 within the same encounter-network.

Formal:

$E_1 \rightarrow E_2$ (causal) iff

StablePattern($C(E_1)$, $C(E_2)$)

across possible mediational variations.

Causation = lawful regularity of constraint transitions,

not hidden powers or noumenal mechanisms.

III.G. Causal Dependence as Cross-Encounter Sequencing

MEOW does not understand causation as the interaction of independently existing substances, nor as the exercise of hidden causal powers. Instead, causal dependence is a feature of the stable sequencing of constraint-patterns within networks of encounter-events. An encounter-event E_1 is causally relevant to E_2 when the constraint-profile of E_1 reliably shapes, limits, or organises the possible configurations of E_2 across variations in mediation. This dependence is neither classical metaphysical necessity nor mere correlation; it reflects the robustness of transition-patterns internal to encounter structure.

Causal asymmetry arises from the organisation of mediation itself. Encounter-events unfold within temporally oriented networks: sensory uptake precedes integrative processing; experimental preparation precedes measurement; environmental change precedes adaptive response. These asymmetries belong to the architecture of encounter, not to substances exerting influence from without. What persists across contexts is not a metaphysical arrow but a directional stability in how encounter-events transition.

This account also accommodates probabilistic causation. Where transitions from E_1 to E_2 show statistical rather than deterministic regularity, causal dependence is expressed as probabilistic constraint: mediation structures the range of possible transitions, and constraint structures their relative likelihoods. Omissions and preventions likewise register as counterfactual disruptions in expected encounter-sequences, rather than as the absence of substance–interaction.

On this view, causation is an emergent property of organised encounter-networks, not a metaphysical glue binding independently existing relata. It explains why scientific modelling identifies robust dependencies without appealing to noumenal mechanisms, and why everyday reasoning tracks causal structure without requiring access to substance-level powers. Causation is a patterned stability in encounter-transitions, not a hidden force behind them.

III.H. Temporal Organisation of Encounter-Events

Encounter-events do not unfold within a pre-existing temporal container. Rather, temporal structure emerges within the organisation of mediation itself. Relations such as simultaneity, succession, and duration are features of how encounters are patterned, not properties imposed from outside the encounter.

Different mediational structures organise temporal sequence differently. Perceptual time includes the "specious present" and the binding of successive inputs into apparent continuity. Biological time expresses developmental rhythms, circadian cycles, and metabolic pacing. Physical time introduces precision via coordinate systems, decay rates, and dynamical equations. These temporal modalities are not competing descriptions of a single temporal substrate; they are distinct modes of temporal organisation within encounter-networks.

Causal asymmetry follows from this organisation. Preparatory encounters constrain subsequent ones in ways that are not symmetrically reversible: sensory uptake precedes integration, environmental change precedes adaptive response, and experimental preparation precedes measurement. The directionality of causation is therefore internal to the mediational architecture rather than evidence for a metaphysical arrow of time.

Past-tense claims are retrodictive structures within present encounters, grounded in the stability of constraint-patterns rather than access to a bygone temporal segment. The persistence of such patterns supports reliable retrodiction without requiring a temporal realism about independent past substances.

IV. Genealogical Positioning

MEOW does not emerge ex nihilo. It stands within a long lineage of relational approaches while rejecting several of their central metaphysical commitments. This section situates MEOW within that broader landscape, tracing affinities and divergences across phenomenology, pragmatism, enactivism, and nondual traditions.

IV.A Phenomenology: Embodiment Without Subject–Object Primacy

Phenomenology—particularly in Merleau-Ponty—recognises that perceptual encounter is not a meeting between a pre-formed subject and a pre-formed world but a *field of embodied engagement*. (Merleau-Ponty, 2012) The body is not an object among objects but the site of orientation, salience, and practical comportment. MEOW inherits the insight that perception is always already structured and that meaning emerges within engagement rather than being added after the fact.

Where MEOW diverges is in its ontological neutrality. Phenomenology often ties encounter to the structure of lived consciousness, whereas MEOW allows the same architectural principles to apply to non-conscious systems. Encounter is relational organisation, whether or not phenomenal character is present. Embodiment matters, but it does not exhaust the domain of encounter-events.

IV.B Pragmatism: Anti-Substantialism Without Behaviourism

Pragmatism—most notably Dewey—rejects the idea that experience can be parsed into a mind confronting an external world (Dewey, 1925). Inquiry, on this view, is a process within an organism–environment circuit. MEOW shares this anti-substantialist orientation and the emphasis on stability emerging from practice rather than from metaphysical substrates.

However, MEOW is not reducible to pragmatism. It does not define truth or meaning solely in terms of practical consequences, nor does it privilege human inquiry as the paradigmatic form of encounter. Instead, it generalises the pragmatic insight: relational organisation is primary, and "mind" and "world" are abstractions derived from stabilised encounter-patterns, not pre-formed domains.

IV.C Enactivism: Participation Without Biological Restriction

Enactivist approaches in cognitive science—especially Varela, Thompson, Rosch (1991)—treat cognition as organism–environment coupling. Experience arises from the dynamic interaction between embodied agents and their worlds. MEOW agrees that cognition is not internal representation but enacted structure, and that mediation and constraint co-constitute the encounter.

Radical enactivist accounts (Di Paolo, Buhrmann, & Barandiaran, 2017) ground encounter in an organism's autonomous activity and sense-making norms. MEOW retains the insight that cognition emerges from organism–environment coupling but rejects the biological restriction: encounter-organisation also occurs in non-autonomous systems such as chemical reactions, tectonic shifts, or computational processes. The architecture of encounter is thus broader than biological sense-making, even though phenomenal presentation remains restricted to reflexively integrated systems.

Where MEOW extends enactivism is in its scope. Enactivism ties encounter to living systems and biological autonomy. MEOW abstracts from this: chemical reactions, tectonic shifts, and computational processes also exhibit encounter-organisation, albeit without phenomenal presentation. The architecture of encounter is broader than biological sense-making.

IV.D Relational and Process Traditions: Relation Without Cosmological Inflation

Process philosophy—especially Whitehead—understands reality as constituted by events rather than substances (Whitehead, 1929). MEOW shares the rejection of substance ontology and the commitment to relational primacy. But it deliberately avoids the cosmological commitments that accompany much of the process tradition: prehension, eternal objects, or universal teleology.

Unlike ontic structural realism (Ladyman & Ross, 2007), MEOW does not posit physical structure as the ontological bedrock. OSR grounds structure in the commitments of fundamental physics, whereas MEOW is structurally pluralistic: encounter-organisation arises in perception, chemistry, geology, computation, and social interaction. MEOW therefore shares OSR's anti-substantivalism but rejects its methodological privileging of physics and its commitment to a single, world-level structural ontology.

MEOW is methodologically modest. It is a theory of the architecture of access, not a metaphysics of cosmic becoming. It draws on the relational orientation of process thought without adopting its speculative machinery. Encounter-events are structural, not metaphysical atoms; their organisation is derived from mediation and constraint, not from metaphysical categories embedded in the universe.

IV.E Agential Realism and Intra-Action: Relata as Emergent Patterns

Barad's agential realism introduces the concept of *intra-action* (Barad, 2007), where relata emerge from relational phenomena rather than pre-existing them. MEOW shares this structural point: the subject–object distinction is derived, not fundamental. Encounter-events precede the abstractions we call mind and world.

The divergence lies in ontological breadth and explanatory minimalism. Agential realism is grounded in quantum measurement and entangles epistemology with an explicitly material ontology. MEOW is intentionally noncommittal about metaphysical materials: encounter-structure arises wherever mediation and constraint cohere, whether in quantum experiments, perception, computation, or chemistry. MEOW offers a broader, more abstract relational ontology.

IV.F Nondual Traditions: Dependence Without Metaphysical Monism

Certain strands of Madhyamaka and Daoist thought deny intrinsic natures and treat relational dependence as fundamental. MEOW echoes this sense of ontological humility: mind and world are not self-standing essences but conceptual abstractions over patterns of relation.

The difference is methodological. Nondual traditions often ground their claims in soteriological or contemplative aims. MEOW, by contrast, develops an analytically tractable account aimed at epistemology, philosophy of science, and the metaphysics of cognition. It inherits the relational insight while shedding the doctrinal frame.

IV.G Summary

MEOW stands within a constellation of relational approaches, yet remains distinct in its commitments:

- It shares phenomenology's attention to structure, but not its subject-centred ontology.
- It shares pragmatism's anti-dualism, but not its behaviourist residues.
- It shares enactivism's emphasis on organisation, but generalises beyond biology.
- It shares process philosophy's relational primacy, but avoids cosmological inflation.
- It shares nondual insights into dependence, but grounds them analytically.

MEOW is thus not a synthesis of these traditions but a reorientation: a minimal relational ontology centred on encounter-events, capable of accommodating mediation and constraint without invoking substances behind experience.

V. What MEOW Is Not

Because the Mediated Encounter Ontology of the World rejects the inherited subject–object architecture, it risks superficial association with several familiar alternatives. This section clarifies those misidentifications. The aim is not polemic but conceptual hygiene: to mark the distinctions that preserve MEOW's minimalism and prevent it from collapsing into neighbouring frameworks.

V.A. MEOW Is Not Subjective Idealism

MEOW does not collapse reality into mind, experience, or representation. Constraint within encounter-events is neither a projection nor a construction of a mental domain. Encounter-structure persists across variations in belief, framing, and conceptual scheme; it is what makes error and correction possible. MEOW therefore rejects the idealist claim that the world is ontologically dependent on consciousness or conceptual activity. Consciousness is one kind of mediational pattern, not the ground of reality.

V.B. MEOW Is Not Physicalist Substantialism

MEOW also rejects substance-based physicalism. Constraint is not evidence for a mind-independent substrate standing "behind" encounter, nor does MEOW posit physical entities as the metaphysical basis of all else. Physical descriptions are aspectual: they capture stable constraint-patterns across encounter-events, not the intrinsic nature of independently existing matter. MEOW allows physical science to describe worldly stability without elevating its categories to ontological primitives.

V.C. MEOW Is Not Neutral Monism

Though MEOW treats "mind" and "world" as derivative abstractions, it is not a neutral monism. There is no primitive neutral "stuff" from which mental and physical emerge. What is primitive is not a substance but a pattern of relational organisation—the co-emergence of mediation, presentational structure, and constraint. MEOW is not a monism of substance but a minimalism of structure.

V.D. MEOW Is Not Kantian Noumenal Dualism

MEOW rejects any appeal to unknowable noumenal substrates. There is no hidden realm of things-in-themselves behind encounter-events. Constraint is internal to encounter-organisation, not a trace or index of something transcendent. Kant's division between phenomena and noumena is replaced by a distinction between what is encountered (structured, mediated, resistive) and unsupported metaphysical projection. The world is not unknowable; it is encounter-structured.

V.E. MEOW Is Not Constructivist Relativism

Because MEOW foregrounds mediation, it may appear sympathetic to constructivism. But constraint prevents collapse into relativism. Mediational differences generate variation, but constraint stabilises encounter-patterns across perspectives, languages, and conceptual schemes. Objectivity, on this view, is not correspondence to a transcendent world but stability across mediational reorganisations. Constructivism overstates the sovereignty of conceptual scheme; MEOW balances mediation with constraint.

V.F. MEOW Is Not Panpsychism or Cosmopsychism

Although MEOW treats presentational structure as a universal feature of encounter-organisation, it does not claim that all encounters possess phenomenal character. Non-conscious encounters (chemical reactions, tectonic shifts, gravitational interactions) exhibit structured mediation and constraint without any experiential dimension. Phenomenal presentation arises only when mediational organisation achieves reflexive integration, characteristic of certain biological systems. MEOW is thus structurally relational, not experientially universal.

V.G Summary

MEOW is not an idealism, physicalism, neutral monism, noumenal dualism, constructivism, or panpsychism. It is a minimal relational ontology anchored in the organisation of encounter-events. This perimeter is crucial: it ensures that MEOW remains a modest, analytically tractable framework rather than a hybrid of pre-existing metaphysical doctrines.

VI. Objections and Responses

No ontology that rejects the subject–object architecture will escape scrutiny from both realist and idealist directions. The following objections are the most likely to arise and illustrate how MEOW addresses them without retreating into the very conceptual frames it replaces.

VI.A Objection: "MEOW is just process philosophy in disguise."

Concern. MEOW's event-relational primacy resembles Whitehead and other process metaphysicians. Critics may argue that "encounter-events" are simply a rebranding of "actual occasions" or that MEOW tacitly inherits process cosmology.

Reply.

MEOW shares with process traditions the rejection of substance ontology, but the similarity ends there. Whiteheadian metaphysics entails:

- prehension as a metaphysical relation,
- concrescence as a universal process,
- eternal objects as invariants of becoming, and
- a cosmological account of the universe's structure.

MEOW adopts none of these. It is deliberately epistemically grounded and metaphysically modest: encounter-events are not metaphysical atoms of the cosmos but the *architecture of access*. MEOW is a theory of how reality is disclosed through mediation and constraint—not a theory of what the universe is "made of." Its commitments remain local, structural, and phenomenologically anchored, not cosmological. To call MEOW "process philosophy" is to overlook its minimalism.

VI.B Objection: "You still owe an account of the substance behind encounters."

Concern. If mediation and constraint co-emerge within encounter-events, what explains the stability of constraint? Doesn't this smuggle in a realist substance or physical ontology behind the event?

Reply.

MEOW does not deny that the world is stable; it denies that stability requires a metaphysical substrate outside encounter. Constraint is a *pattern across encounters*, not a power emanating from a hidden realm. For example:

- gravitational regularities,
- chemical reaction pathways,
- geometric invariances,
- persistence of perceptual illusions

all appear as stable constraint-patterns in actual encounters. No explanatory work is done by positing a substrate behind those patterns. MEOW treats the stability itself—the recurrent organisation of constraint—as ontologically primary. The demand for a substance behind that stability is a residue of the subject–object architecture that MEOW rejects.

VI.C Objection: "If encounter-events require mediation, how do you explain dinosaurs?"

Concern. A classic realist challenge: if encounter-events presuppose mediation, how did reality function before humans or conscious organisms? Did dinosaurs exist only when observed?

Reply.

MEOW does *not* claim that encounter-events require consciousness. Non-conscious systems—tectonic plates, radioactive decay, molecular interactions—have presentational structure without phenomenal presentation. They are encounter-events organised by mediation at chemical, physical, and dynamical levels.

Past-tense claims ("dinosaurs existed 65 million years ago") are grounded in constraint patterns of present encounters—fossils, isotopic decay curves, geological stratification. These constraints support retrodictive inference within encounter. MEOW does not privilege human consciousness; it generalises the architecture of encounter across conscious and non-conscious systems. Dinosaur reality is inferred from present constraints, not accessed by stepping outside encounter.

VI.D Objection: "MEOW eliminates mind-independent reality."

Concern. Without a notion of an external world, realism fears that MEOW collapses into constructivism or idealism.

Reply.

MEOW eliminates the *substance* notion of mind-independent reality, not the stability that realism is rightly concerned with. Constraint-patterns are independent of the particularities of any conscious agent's framing, belief, or conceptual apparatus. Gravity does not care about interpretation; decay constants are not culturally relative; perceptual illusions persist despite knowledge.

Objectivity in MEOW is:

stability across mediational variation,

not access to a noumenal world.

This provides all the empirical independence that scientific realism requires without presupposing a hidden substance outside encounter.

VI.E Objection: "MEOW collapses into panpsychism."

Concern. If every encounter has presentational structure, critics may infer that MEOW smuggles in proto-experience everywhere.

Reply.

MEOW distinguishes presentational structure from phenomenal presentation. All encounters have organisation; only some have experience. Chemical reactions have mediation and constraint; they do *not* have phenomenal character. Consciousness arises only when mediational architectures achieve reflexive integration (as in certain biological systems). MEOW is relational, not experientially universal.

VI.F Objection: "This does not solve causation, only redescribes it."

Concern. Critics may argue that describing causal regularity in terms of encounter-transition is insufficient.

Reply.

MEOW does not seek to discover hidden causal powers. It clarifies the structure that underlies causal inference:

- stable constraint transitions across encounter-events,
- directional organisation of mediation,
- probabilistic constraint where determinism fails,
- counterfactual sensitivity understood as disruption of expected sequences.

This is precisely what scientific practice relies on. MEOW rejects metaphysical inflation in favour of structural articulation. It provides a minimal ontology that supports causal explanation without positing noumenal mechanisms.

VI.G Summary

The objections converge on a single theme: a desire to preserve the subject–object metaphysics MEOW rejects. MEOW's replies demonstrate that:

- no hidden substances are required,
- no appeal to consciousness is necessary,
- scientific objectivity is preserved,
- realism and idealism both overreach,
- relational structure suffices for epistemic, causal, and ontological purposes.

MEOW's novelty lies not in inventing new metaphysical furniture but in clearing away the furniture that obscured the architecture of encounter itself.

VII. Philosophical Consequences and Future Directions

The Mediated Encounter Ontology of the World reconfigures several longstanding philosophical debates by reframing their foundational assumptions. Rather than adjudicating between competing metaphysical pictures, MEOW dissolves the inherited subject–object architecture and redistributes explanatory burdens across the structure of encounter-events. This section sketches several of the most salient consequences and indicates directions for future development.

VII.A Epistemology: Objectivity Without Noumena

MEOW reframes epistemic justification by rejecting the dichotomy between unmediated access and subjective construction. Objectivity is not correspondence to a metaphysical exterior but stability across mediational variation. Error and correction arise when expected encounter-organisation diverges from actual constraint-patterns. Because constraint is internal to encounter, epistemic reliability is grounded not in access to an external world but in the convergence of differently mediated encounters.

This approach supports a form of encounter-coherent realism: claims are warranted when they maintain coherence across shifts in perspective, instrumentation, or conceptual framing. Foundationalism and coherentism both lose their central tension—foundation vs. system—since encounter-events already integrate mediation and constraint.

VII.B Philosophy of Science: Laws as Constraint-Patterns

Scientific practice systematically identifies stable constraint-patterns across encounter-events. MEOW therefore interprets scientific laws not as descriptions of noumenal necessity but as codifications of:

- cross-contextual stability,
- recurrent structural organisation, and
- reliable transition-patterns within encounter-networks.

Theory change becomes intelligible as the refinement of mediational organisation (new instruments, models, computational tools) that yields access to more stable or discriminating constraint-patterns. Underdetermination is tempered by the fact that convergence across diverse mediational pathways constrains theoretical arbitrariness.

Scientific realism is thereby reconstructed without metaphysical overreach: what science tracks is constraint-structure, not substances behind experience.

VII.C Consciousness and Embodiment: Presentation Without Dualism

MEOW offers a minimal account of consciousness as a special case of presentational structure. Phenomenal presentation arises when mediational patterns achieve reflexive integration—when an encounter becomes aware of its own organisation. This bypasses the "hard problem," which presupposes a substance dualism (physical vs. mental) that MEOW rejects.

Embodiment remains central: sensorimotor mediation grounds how encounters are structured. But consciousness is neither the foundation of reality nor an inexplicable add-on. It is one mode of organisation among others.

This provides a non-reductive, non-dualistic account of consciousness suitable for integration with biological, computational, or enactive models.

VII.D Extended and Enactive Cognition: No Boundary Problem

Traditional debates over the "mark of the cognitive" ask where cognition is located—inside the skull or extended into tools, bodies, and environments. MEOW sidesteps this by treating the boundaries of encounter-events as organisational, not metaphysical. If the mediational–constraining organisation that constitutes a cognitive encounter includes notebooks, instruments, or social scaffolding, then these are not "extensions" but components of the encounter itself.

This reframes cognitive extension as a question about the structure of encounter-events rather than about the location of mental substance. The extended mind debate becomes a special case of encounter-organisation.

VII.E Quantum Measurement and Relational Context

Although MEOW does not commit to any particular interpretation of quantum mechanics, it provides conceptual resources for understanding measurement without invoking classical subject–object metaphysics. Quantum measurement can be treated as an encounter-event: mediation = apparatus and interaction-framework; constraint = stable statistical patterns; presentational structure = the recorded outcomes.

MEOW thereby aligns with relational or contextual interpretations while avoiding their tendency toward metaphysical inflation. The measurement problem becomes a problem of encounter-organisation, not of ontological collapse.

VII.F Artificial and Synthetic Minds: Criteria Without Anthropocentrism

MEOW offers a framework for evaluating non-biological systems without reducing cognition to computation or tying it to human-like consciousness. The relevant question becomes:

Does the system exhibit reflexively integrated mediational organisation capable of generating phenomenal presentation?

This shifts the discourse from substrate-based criteria ("silicon vs. carbon") to organisational criteria grounded in encounter-structure. MEOW provides a principled way to assess artificial minds without anthropocentric bias or simplistic functionalism.

VII.G Future Work

MEOW opens several pathways for further development:

- Temporal organisation of encounters: articulating how temporality arises from encounter-sequences.
- Compositionality: analysing how simple encounters combine into complex structures.
- Normativity: exploring how evaluative and ethical patterns emerge from mediated interaction.
- Formal dynamics: developing mathematical or computational models of encounter-organisation.
- Domain-specific extensions: neuroscience, AI, quantum theory, and social ontology.

These developments are not additions to MEOW's metaphysical core but elaborations of its structural insights. MEOW's account of encounter-structure also provides resources for future analysis in areas such as cognitive science, extended cognition, and quantum measurement, though these lie beyond the scope of the present paper.

VII.H Summary

MEOW's consequences are not speculative add-ons but natural extensions of its central claim: encounter-events are ontologically basic. Once the subject–object architecture is relinquished, longstanding philosophical problems take on a different shape—less about metaphysical discovery, more about structural articulation. The framework provides a unified lens through which epistemic, scientific, cognitive, and metaphysical questions can be reframed without appealing to substances behind experience or minds beyond encounter.

VIII. Conclusion

The Mediated Encounter Ontology of the World begins from a simple diagnostic insight: the realism–idealism debate persists not because one side has failed to overcome the other, but because both inherit an ontological architecture that experience does not vindicate. By presupposing that "mind" and "world" are independently existing substances that later enter into epistemic relation, the debate generates a set of impossible demands—unmediated access for realism, unconstrained construction for idealism—that no account can satisfy without contradiction.

MEOW resolves this not by choosing between these positions but by rejecting the shared premise that sustains them. Encounter-events, not substances, are ontologically primary. Mediation, presentational structure, and constraint arise together within the organisation of encounter. "Mind" and "world" are abstraction-patterns distilled from regularities across such events, not metaphysical domains awaiting reconciliation. Objectivity is stability across mediational variation, not correspondence to an inaccessible substrate. Causation is the patterned sequencing of constraint, not the action of hidden powers.

By reframing the architecture of access rather than defending a substantive worldview, MEOW dissolves the inherited dichotomies—inner and outer, mental and physical, appearance and reality—that have long structured philosophical disputes. It provides a minimal relational ontology capable of grounding epistemic reliability, scientific practice, temporal and modal inference, and a principled account of consciousness without invoking the substances or dualisms that generated those problems in the first place.

What remains is not a final metaphysical picture but an open framework. MEOW does not close inquiry; it clears conceptual space. Future work may extend its account of temporal organisation, refine its treatment of complex encounter-networks, or develop formal models of encounter-structure. The aim is not to arrest metaphysical imagination but to redirect it: away from the search for entities behind experience and toward the articulation of the structures through which reality is disclosed.

If the subject–object architecture obscured more than it revealed, MEOW offers a way to set it aside. What it replaces it with is not a new metaphysical substance but a more faithful account of how we—biological, cognitive, cultural, technological beings—actually meet the world. Subsequent work will develop the temporal, normative, and scientific implications of MEOW, extending the framework without altering its core relational commitments.

References

Adams, F., & Aizawa, K. (2001). *The bounds of cognition. Philosophical Psychology,* 14(1), 43–64. https://doi.org/10.1002/9781444391718.ch4

Barad, K. (2007). *Meeting the universe halfway: Quantum physics and the entanglement of matter and meaning*. Duke University Press.

Berlin, B., & Kay, P. (1969). *Basic color terms: Their universality and evolution.* University of California Press.

Clark, A. (2013). Whatever next? Predictive brains, situated agents, and the future of cognitive science. *Behavioral and Brain Sciences, 36*(3), 181–204. https://doi.org/10.1017/S0140525X12000477

Clark, A. (2016). *Surfing uncertainty: Prediction, action, and the embodied mind.* Oxford University Press.

Clark, A., & Chalmers, D. (1998). The extended mind. *Analysis, 58*(1), 7–19.

Dalrymple, G. B. (1991). *The age of the Earth*. Stanford University Press.

Dewey, J. (1925). *Experience and nature*. Open Court.

Di Paolo, E., Buhrmann, T., & Barandiaran, X. (2017). *Sensorimotor life: An enactive approach to the human condition*. Oxford University Press.

Friston, K. (2010). The free-energy principle: A unified brain theory? *Nature Reviews Neuroscience, 11*(2), 127–138. https://doi.org/10.1038/nrn2787

Hoffman, D. D. (2019). *The case against reality: Why evolution hid the truth from our eyes*. W. W. Norton.

Hohwy, J. (2013). *The predictive mind*. Oxford University Press.

Ladyman, J., & Ross, D. (2007). *Every thing must go: Metaphysics naturalized.* Oxford University Press.

Levinson, S. C. (2003). *Space in language and cognition: Explorations in cognitive diversity*. Cambridge University Press.

Merleau-Ponty, M. (2012). *Phenomenology of perception* (D. A. Landes, Trans.). Routledge. (Original work published 1945)

Roberson, D., Davies, I., & Davidoff, J. (2000). Color categories are not universal: Replications and new evidence from a stone-age culture. *Journal of Experimental Psychology: General, 129*(3), 369–398. https://psycnet.apa.org/doi/10.1037/0096-3445.129.3.369

Segall, M. H., Campbell, D. T., & Herskovits, M. J. (1966). *The influence of culture on visual perception*. Bobbs-Merrill.

Uexküll, J. von (2010). *A foray into the worlds of animals and humans: With a theory of meaning* (J. D. O'Neil, Trans.). University of Minnesota Press. (Original work published 1909)

Varela, F. J., Thompson, E., & Rosch, E. (1991). *The embodied mind: Cognitive science and human experience*. MIT Press.

Whitehead, A. N. (1929). *Process and reality: An essay in cosmology*. Macmillan.

Whorf, B. L. (1956). *Language, thought, and reality: Selected writings of Benjamin Lee Whorf* (J. B. Carroll, Ed.). MIT Press.

Further Reading

(Recommended sources that situate MEOW within broader philosophical, scientific, and historical contexts. Not cited directly.)

Process Philosophy / Relational Metaphysics

James, W. (1912). *Essays in radical empiricism*. Longmans, Green & Co.

Stengers, I. (2011). *Thinking with Whitehead: A free and wild creation of concepts* (M. Chase, Trans.). Harvard University Press. (Original work published 2002)

Phenomenology / Embodiment

Gallagher, S. (2017). *Enactivist interventions: Rethinking the mind*. Oxford University Press.

Thompson, E. (2010). *Mind in life: Biology, phenomenology, and the sciences of mind.* Harvard University Press.

Enactivism / 4E Cognition

Hutto, D. D., & Myin, E. (2013). *Radicalizing enactivism: Basic minds without content.* MIT Press.

Rowlands, M. (2010). *The new science of the mind: From extended mind to embodied phenomenology.* MIT Press.

Constructivism / Anti-Realism

Goodman, N. (1978). *Ways of worldmaking.* Hackett.

Putnam, H. (1981). *Reason, truth, and history.* Cambridge University Press.

Rorty, R. (1979). *Philosophy and the mirror of nature.* Princeton University Press.

Relational / Quantum Approaches

Bohr, N. (1935). Can quantum-mechanical description of physical reality be considered complete? *Physical Review, 48*(8), 696–702.

Rovelli, C. (1996). Relational quantum mechanics. *International Journal of Theoretical Physics, 35*(8), 1637–1678.

Panpsychism / Boundary-Setting

Goff, P. (2017). *Consciousness and fundamental reality.* Oxford University Press.

Strawson, G. (2006). Realistic monism: Why physicalism entails panpsychism. *Journal of Consciousness Studies, 13*(10–11), 3–31.

AI, Synthetic Minds, and Agency

Botvinick, M., Barrett, H. C., & Griffiths, T. L. (2023). Machine behaviour. *Nature Reviews Neuroscience, 24*, 535–548.

Dennett, D. C. (1991). *Consciousness explained.* Little, Brown.

Glossary of Terms

Terms are defined as they function within the Mediated Encounter Ontology of the World. Where a term has a standard philosophical usage that the framework repurposes, the departure is noted. Section references indicate where the term is introduced or most fully developed.

Affordance. An organism–field compatibility gradient constituted by encounter. Affordances arise when a self-maintaining system differentiates the constraint-field into what supports and what obstructs its persistence. They are relational, continuous, and context-sensitive – neither objective properties of the environment nor subjective projections of the organism. Adapted from Gibson (1979) but reframed: affordances are encounter-constituted rather than objectively present in the environment. (§6.2)

Apophatic mind. The diagnostic observation that contemporary discourse about AI and mind operates through successive negation – defining 'mind' by what machines cannot do rather than by what mind demonstrably is. The term draws on the apophatic (via negativa) tradition in theology, in which the object of inquiry is approached through what it is not. The result is a concept whose content shrinks as artificial capacities expand, sustained by grammatical persistence rather than positive specification. (§10.2)

Aspectual designation. The classification of encounter-events as 'mental' or 'physical' by emphasis rather than by ontological kind. Within MEOW, 'mental' and 'physical' are not names for distinct substances but different ways of describing the organisation of the same relational event. (§1.3)

Bandwidth problem. The structural limitation inherent in any finite encoding system. Language, as a synchronisation protocol, has finite resolution; as conceptual complexity increases, the protocol's capacity to synchronise stabilisations across agents degrades. The bandwidth problem is not a defect to be remedied by better vocabulary but a consequence of the protocol's operating conditions. (§7.7)

Coherence grammar. The account of logic developed within MEOW. Logic is neither Platonic (existing independently in an abstract realm) nor conventional (adopted by agreement). It is the minimal structural condition required to preserve consistency in stabilised orientation under constraint. Logical principles are stability conditions extracted from the operating requirements of any system that must maintain coherent orientation within a constraint-field. (§4.2)

Coherence zone. A region of the constraint-field within which a set of stabilisations can be maintained without internal contradiction. A formal term of art designating the domain of applicability of a given coherence grammar. Different coherence grammars (classical, intuitionistic, paraconsistent) define different coherence zones. (§4.2–4.4)

Compression. The process by which a modally differentiated system selects, abstracts, categorises, and stabilises encounter-patterns into workable representations. All knowledge, all language, and all institutional categories involve compression. Compression introduces structural loss: no stabilisation captures the full constraint-structure of the domain it represents. (§5.3)

Constraint. The dimension of an encounter-event that is not freely variable – the patterned invariance, pushback, or recalcitrance that persists across variation in mediation. Constraint is not imported from a noumenal realm, nor projected by a subjective mind. It arises from the interplay between mediational architecture and the structural regularities that persist across encounters. Constraint is MEOW's realist anchor. (§1.5, 2.3)

Constraint-cluster. A dense, relatively stable grouping of constraint-patterns at a particular scale of encounter. Objects, in the MEOW framework, are scale-dependent constraint-clusters. Legal and institutional categories are constraint-clusters maintained through second-order coordination. (§2.6, 8.6)

Constraint-field. The structured domain of patterned recalcitrance within which encounter-events occur and stabilisations are tested. The constraint-field is not a background container or hidden substrate, but the relational field of non-freely-variable structure as it bears upon mediation. Different mediational architectures organise the same constraint-field differently, but they do not fabricate the constraint-patterns that persist across variation. (§1.5, 2.3, 5.2)

Constraint transmission. The mechanism of causation within MEOW. An encounter-event E_1 is causally relevant to E_2 when the constraint-profile of E_1 reliably shapes the possible configurations of E_2 across mediational variations. Causation is not substance-level production but the stability of constraint propagation across encounter-sequences. (§2.9)

Convergence realism. MEOW's form of realism. Reality is grounded not in substance-ontological independence but in the convergent stability of relational constraint across distinct mediational architectures. When differently mediated encounter-systems converge on the same constraint-patterns, that convergence is evidence of invariant structure rather than coincidence of projection. Distinguished from naïve object realism, scientific realism, constructivism, Kantian noumenalism, and relativism. (§9.1–9.3)

Convergent stability. The operative meaning of objectivity within MEOW. A claim is objective to the extent that the constraint-pattern it tracks is stable across encounters organised by different mediational architectures. Objectivity admits of degrees: strong in the Invariants region, weaker in the Contestables and Fluids regions. Replaces the 'view from nowhere' as the criterion of objectivity. (§9.4)

Cross-perspectival convergence. The central evidential mechanism for convergence realism. When distinct mediational architectures – different species, different instruments, different conceptual frameworks – converge on the same constraint-patterns, the convergence is more parsimoniously explained by shared constraint than by coincidence of fabrication. The more the mediational architectures diverge, the stronger the evidential force. (§9.3)

Diachronic narrative compression. The mechanism through which the 'self' is constituted. A stabilised pattern tracking organisational continuity across temporally extended sequences of encounter-events. The self is not a substance that persists through time but a heuristic compression of modal differentiation across a temporal trajectory. Useful for social coordination; misleading when mistaken for a metaphysical entity. (§3.5)

Diagnostic method. The framework's characteristic mode of engagement with philosophical problems. Rather than answering inherited questions on their own terms, the method traces problems to the structural misdrawings that produce them – typically to substance-ontological assumptions, ontological grammar effects, or the conflation of encoding artefacts with features of reality. (Preface)

Double-entry ledger. The transfer mechanism through which the apophatic dynamic operates in AI–mind discourse. Capacities previously attributed to mind are credited to mechanism as machines demonstrate them; the 'mind' column shrinks correspondingly. Structurally identical to the historical transfer between religion and science. (§10.3)

Effectiveness gradient. The Language Effectiveness–Complexity Gradient (Figure 7.1). A representation of the relationship between conceptual complexity and linguistic effectiveness, showing how the synchronisation protocol's performance degrades as complexity increases. Four regions are marked: Invariants (high effectiveness), Contestables (degrading), Fluids (manifestly inadequate), and Ineffables (structural limit). The shaded region between the actual and presumed effectiveness curves is the overestimation gap. (§7.8–7.9)

Encounter-event. The ontological primitive of the MEOW framework. A structured relational occurrence in which three aspects co-emerge: mediation (the organisational structure through which the encounter arises), presentational structure (the pattern through which it is manifested), and constraint (the dimension that is not freely variable). Encounter-events are not relations between antecedently existing relata; they are the primitive from which relata are retrospectively abstracted. (§1.3)

Encounter-organisation. The structural arrangement of an encounter-event – the way mediation, presentational structure, and constraint are configured within it. Different mediational architectures produce different encounter-organisations of the same constraint-field. Used throughout the monograph as the general term for how encounters are patterned.

Encounter pressure. The ongoing testing of stabilisations by the constraint-field. Every prediction that fails, every affordance that misfires, every expectation that collapses is an instance of encounter pressure. Stabilisations that survive encounter pressure persist; those that do not are revised or abandoned. The epistemic selection mechanism through which knowledge emerges. (§5.2)

False convergence. The appearance of shared meaning where underlying stabilisations are structurally incommensurate. Occurs when the synchronisation protocol transmits words successfully at the syntactic level whilst the stabilisations reconstructed at the receiving end diverge from those compressed at the sending end. Characteristic of the Contestables region of the effectiveness gradient and a primary mechanism of institutional dysfunction. (§7.9, 8.7)

Filtering does not imply fabrication. The principle that mediational shaping of encounter does not manufacture the constraint-patterns that persist across mediational variation. MEOW's realist hinge: mediation determines the form of an encounter; constraint determines what is stable within it. The full abductive defence is provided through the convergence argument. (§2.8, 9.7)

Fluency tax. The asymmetry in articulatory cost between substance-ontological and relational-ontological formulations in noun-dominant grammar. Substance ontology flows with Indo-European subject-predicate structure; relational ontology works against it. The framework accepts this tax deliberately as the cost of structural fidelity over cognitive economy. (§7.10)

Four-tier claim classification. The epistemic-status framework used in the Preface to sort the monograph's claims into four categories: inherited empirical premises (drawn from established science), architectural proposals (the framework's own structural commitments), diagnostic applications (consequences of the architecture applied to inherited problems), and open frontiers (questions the framework identifies but does not resolve). (Preface)

Heuristic container. A label that bundles together a cluster of organisationally diverse phenomena under a single noun because coordination requires it. 'Mind' is the paradigmatic example: it bundles consciousness, agency, intention, reflexivity, and phenomenal experience into a single word. The container persists because it solves a recurrent coordination problem, not because it names a natural kind. (§10.8)

Indexical compression. The mechanism through which the first-person pronoun 'I' is constituted. A linguistic device that compresses the entire modal differentiation of a self-maintaining process-cluster into a single referential token. Simultaneously indispensable (as a coordination device) and misleading (as an ontological signal), because its grammatical role as a noun reinstalls the substance architecture the framework rejects. (§3.4)

Invariant anchor. Relational constraint as the ground of MEOW's realism. What persists across encounters is not substance but patterned invariance – structural regularities that remain stable across variations in mediation. The invariant anchor replaces substance-ontological mind-independence as the criterion of reality. (§2.3)

Language Insufficiency Hypothesis (LIH). The structural claim that for any finite-bandwidth linguistic protocol operating across non-identical mediational architectures, there exist domains in which the constraint-patterns of encounter resist stable compression with sufficient fidelity to achieve reliable synchronisation. In less formal terms: there are things that language cannot say – not because speakers are inarticulate, but because the encoding protocol's resolution is insufficient for the domain. Developed independently in Willis (2026) and restated as a consequence of MEOW's architecture. (§7.8)

Meaning. Salience-weighted affordance structure relative to organismal persistence. Meaning is pre-linguistic: it arises in the encounter between a self-maintaining system and the constraint-field before any linguistic encoding occurs. Language compresses and synchronises pre-existing meaning-structures across agents, at the cost of fidelity. (§6.3)

Mediation. The structural condition under which any encounter-event arises. Not an epistemic obstacle but a constitutive feature of encounter. Triply layered: biological (species-specific sensory and motor organisation), cognitive (predictive and inferential structuring), and linguistic–cultural (inherited schemes of classification, interpretation, and valuation). Mediation is ontological rather than merely epistemic: it is part of what reality is, not an obstacle to knowing it. **(§1.4)**

MEOW (Mediated Encounter Ontology of the World). The relational metaphysics developed in this monograph. MEOW takes encounter-events as ontologically primary, treats 'mind' and 'world' as derivative abstractions, and locates reality's invariant structure in relational constraint rather than in substance. Originally proposed in Willis (2025); extended across eleven chapters in the present work. (§1.1–1.3)

Modal differentiation. The account of subjecthood within MEOW. The point at which the interactional field becomes differentiated into constraint and affordance relative to a self-maintaining, constraint-responsive process-cluster. The subject is not outside the field; it is a dynamic differentiation within it. The formulation is substrate-neutral: biological organisms are the paradigmatic but not sole possible locus. (§3.2)

Normativity frontier. The boundary at which the architecture's descriptive resources reach their limit with respect to moral obligation. Identified and dissolved in Chapter 11 through the adoption of moral non-cognitivism: normative discourse is socially real (it constrains encounter and alters interactional trajectories) but not ontologically primitive (it does not track mind-independent moral facts). (§11.1–11.7)

Ontological grammar. The structural commitments imposed on encoded content by the grammatical architecture of natural language. Indo-European subject-predicate grammar scaffolds substance ontology by making object-first thinking fluent and relational thinking cumbersome. The ontological grammar thesis is not linguistic determinism: speakers can think against their grammars, but the grammar taxes alternatives, and cognitive economy favours the path of least resistance. (§7.3–7.6)

Organismal persistence. The continued maintenance of a self-maintaining process-cluster across time. Within MEOW, persistence is not mere duration but the ongoing preservation of organisational coherence under perturbation. Affordance, meaning, salience, and valence are all structured relative to what supports, threatens, or reorganises organismal persistence. (§3.3, 6.2–6.3)

Overestimation gap. The domain between the actual effectiveness curve and the presumed effectiveness curve on the Language Effectiveness–Complexity Gradient (Figure 7.1). The region in which language is assumed to function but structurally cannot. Where philosophical impasse, institutional dysfunction, and the reification of heuristic categories most characteristically occur. (§7.8, 8.8)

Phenomenal presentation. A special case of presentational structure. The experiential character that arises when mediational patterns achieve reflexive integration – when the encounter incorporates its own prior states, differentiates self from other, and integrates sensory, affective, and cognitive signals into a unified field of experience. Not all encounter-events have phenomenal presentation; the framework is relational, not panpsychist. (§1.6)

Presentational structure. The organised pattern of any encounter-event, conscious or otherwise. The way the encounter is manifested – its structural profile, its internal organisation, the form in which it appears or could appear to an appropriate mediational system. All encounter-events possess presentational structure; phenomenal presentation is the special case that arises under reflexive integration. (§1.6)

Reification error. See apophatic mind (§10.5).

Reflexive integration. The organisational achievement by which an encounter-system incorporates its own prior states into present activity, differentiates self from other, and integrates sensory, affective, and cognitive signals into a unified experiential field. Within MEOW, phenomenal presentation arises when presentational structure reaches this level of reflexive integration. (§1.6, 3.5)

Salience. The differential weighting by which aspects of an encounter stand out for a self-maintaining system relative to its organisation, history, and current conditions. Salience is not merely attentional prominence but organism-relative significance within the affordance structure of the encounter. It helps determine which features are stabilised, acted upon, or ignored. (§6.3)

Scale-dependence. The property of objects within MEOW. Objects are stabilisations of relational constraint-patterns relative to a particular scale of encounter determined by the mediational architecture of the encountering system. A boulder is a stable constraint-cluster at the human scale; at the particle scale it dissolves; at the geological scale it flows. Scale-dependence is not a defect in the object's reality but a feature of what it means for reality to be structured interaction rather than substance. (§2.6)

Second-order constraint system. An institution. A pattern of stability constituted and maintained through linguistic coordination and sustained by coordinated behaviour, enforcement mechanisms, and material embedding. Distinguished from first-order constraints (physical, perceptual, cognitive) by dependence on the synchronisation protocol for their stability. Simultaneously powerful (language enables coordination at scales no other mechanism can achieve) and fragile (the protocol has structural limits the institution cannot acknowledge without undermining its own authority). (§8.3)

Self-maintenance. The organisational feature that gives rise to modal differentiation. A self-maintaining process-cluster actively preserves its own structural coherence against perturbation, generating a differential relationship to its environment in which some interactions support and others threaten its persistence. The origin of perspective within MEOW: the subject does not have a perspective; the subject is the perspective. (§3.3)

Stabilisation. The process by which a modally differentiated system forms a durable orientation within the constraint-field. Knowledge is stabilisation under encounter pressure. Truth is the long-term survivor of encounter pressure across populations of agents. Stabilisations are always compressed, always partial, and always subject to revision. (§5.1–5.2)

Substance ontology. The metaphysical framework that MEOW rejects. The view that reality consists of independently existing things (substances) that bear properties and enter into relations. Critiqued throughout the monograph as generating pseudo-problems – the mind–world bifurcation, the no-relata demand, the hard problem of consciousness – that dissolve when the substance assumption is withdrawn. Its persistence is partly explained by the ontological grammar of Indo-European languages. (§1.1, 2.7, 7.3)

Synchronisation protocol. Language, reconceived. A finite-bandwidth mechanism for aligning predictive stabilisations across distinct mediational architectures. Language does not name or represent; it coordinates. Communication is approximate synchronisation under constraint, not duplication of inner representations. The protocol has structural limits (the bandwidth problem) and structural biases (ontological grammar). (§7.2)

Synchronisation repair. The mechanisms through which institutions patch the gap between what their linguistic categories say and what the constraint-patterns require. Includes precedent, authoritative reinterpretation, and strategic ambiguity. These mechanisms are the immune system of linguistically constituted coordination, not symptoms of institutional weakness. (§8.5)

Valence. The positive, negative, or threat-relevant loading of an encounter-pattern relative to organismal persistence. Valence is not an intrinsic property of an object but a relational feature of how an affordance bears upon self-maintenance. It marks not what an encounter is in itself, but how it matters for the continued organisation of the encountering system. (§3.3, 6.2–6.3)

Vulnerability scan. The editorial process through which the manuscript's claims were tested for overcommitment, inconsistency, and exposure to foreseeable objections. Conducted across multiple AI systems in parallel, with the author adjudicating feedback. Key findings included the removal of the Hopi linguistic example, moderation of cross-linguistic overclaiming, and footnoting of the powers metaphysics alternative.

Index

K

L

M

N

O

P

Q

R

S

T

V

W

About the Author

Bry Willis is an independent philosopher whose work explores the cracks in language, objectivity, and meaning. He writes at the intersection of philosophy of language, meta-philosophy, and social epistemology, often testing how concepts fracture under pressure. His essays – such as *Objectivity Is Illusion: An Operating Model of Social and Moral Reasoning* – trace how consensus masquerades as truth and how rhetoric, power, and perspective shape what we call knowledge. He manages the *Philosophics Blog* and publishes literary speculative non-fiction under the pseudonym Ridley Park.

Other Titles by Bry Willis

Books:

A Language Insufficiency Hypothesis: Mapping the Boundaries of Linguistic Expression
ISBN: 978-0-9710869-4-4

When Language Fails: Ontological Pluralism and the Limits of Moral Resolution
ISBN: 978-0-9710869-4-4

Illusions of Light: Thinking After the Enlightenment
ISBN: 978-0-9710869-7-5

Meditations on Nothing: Notes Before Existence
ISBN: 978-0-9710869-5-1

Meditations on Nothing: A Companion Guide
ISBN: 978-0-9710869-6-8

Essays by Bry Willis

Essays are available as open-access publications through Zenodo.

The Anti-Enlightenment Essays:

Against Agency: The Fiction of the Autonomous Self

Competency, Proxies, and Political Standing: A Conceptual Diagnosis or On the Rhetoric of Democratic Inclusion

Moral Universality and Its Discontents: A Critical Examination of Normative Ethics' Conceptual Foundations

Objectivity Is Illusion: An Operating Model of Social and Moral Reasoning
Rational Ghosts: Why Enlightenment Democracy Was Built to Fail
Temporal Ghosts: The Tyranny of the Present
The Discipline of Dis-Integration: Philosophy Without Redemption
The Fact of the Matter: After Bernard Williams – Truthfulness, Facts, and the Myth of Immediacy
The Great Substitution: From Metaphysics to Metaphysics The Grammar of Impasse – Causal Mislocation
The Grammar of Impasse – Conceptual Exhaustion The Myth of Homo Normalis: Archaeology of the Legible Human

The Will to Be Ruled: Totalitarianism and the Fantasy of Freedom

Other Essays:

Disagreement Without Referees: Ontological Incommensurability and the Limits of Moral Adjudication
Grammatical Failure: Why Liberal Epistemology Cannot Diagnose Indoctrination
Illegible Futures: Affective Legibility and the Governance of Optimism
Language As Interface: Underconstraint, Genealogy, and Moral Incommensurability
Quand le langage fonctionne sans trancher : Indétermination structurelle et termes normatifs en contexte pluraliste Seeing Like a System: High Modernism, Legibility, and the Limits of Intelligent Robotics
The Mediated Encounter Ontology of the World: A Relational Metaphysics Beyond Mind and World
The Expiration Date of Object Permanence: Heuristics, Grammar, and Quantum Pseudoproblems
Truth After Deflation: Why Truth Resists Stabilisation
The Rhetoric of Evil: How a Theological Artefact Survived Secular Moral Thought
When Syntax Is Asked to Bear Too Much: The Frege–Geach Problem and the Limits of Linguistic Invariance

Connect

Further work: philosophics.blog

Inquiries: talk@philosophics.blog *or* microglyphics@gmail.com

Philosophics Press

Philosophics Press publishes works that inhabit the fault lines between philosophy, literature, and design. Founded as an imprint of Microglyphics, it exists to sustain thought outside the economies of visibility and prestige – to print, quite literally, what persists when systems fail.

Made in the USA
Las Vegas, NV
27 July 2026

51812982R00164